New Frontiers in Robotics — (Vol. 3)

SIMULTANEOUS LOCALIZATION AND MAPPING

Exactly Sparse Information Filters

NEW FRONTIERS IN ROBOTICS

Series Editors: Miomir Vukobratovic *("Mihajlo Pupin" Institute, Serbia)*
Ming Xie *(Nanyang Technological University, Singapore)*

New Frontiers in Robotics — 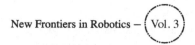 Vol. 3

SIMULTANEOUS LOCALIZATION AND MAPPING

Exactly Sparse Information Filters

Zhan Wang · Shoudong Huang · Gamini Dissanayake

University of Technology, Sydney

Australia

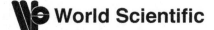

NEW JERSEY · LONDON · SINGAPORE · BEIJING · SHANGHAI · HONG KONG · TAIPEI · CHENNAI

Published by

World Scientific Publishing Co. Pte. Ltd.
5 Toh Tuck Link, Singapore 596224
USA office: 27 Warren Street, Suite 401-402, Hackensack, NJ 07601
UK office: 57 Shelton Street, Covent Garden, London WC2H 9HE

British Library Cataloguing-in-Publication Data
A catalogue record for this book is available from the British Library.

New Frontiers in Robotics — Vol. 3
SIMULTANEOUS LOCALIZATION AND MAPPING
Exactly Sparse Information Filters

ISBN-13 978-981-4350-31-0
ISBN-10 981-4350-31-1

Printed in Singapore.

Preface

Autonomous vehicles, such as mobile robots, are required to navigate intelligently in the scenarios of indoor guidance, robot rescue, fire fighting, underwater exploration etc. Efficiently acquiring a geometric representation of the environment is crucial in the success of these tasks.

This book is mainly concerned with computationally efficient solutions to the feature-based simultaneous localization and mapping (SLAM) problem. The setting for the SLAM problem is that of a robot with a known kinematic model, equipped with on-board sensors, moving through an environment consisting of a population of features. The objective of the SLAM problem is to estimate the position and orientation of the robot together with the locations of all the features.

Extended Kalman Filter (EKF) based SLAM solutions widely discussed in the literature require the maintenance of a large and dense covariance matrix when the number of features present in the environment is large. Recently, Extended Information Filter (EIF) based SLAM solutions have attracted significant attention due to the recognition that the associated information matrix can be made sparse, which leads to significant computational savings. However, existing EIF-based SLAM algorithms have a number of disadvantages, such as estimator inconsistency, a long state vector or information loss, as a consequence of the strategies used for achieving the sparseness of the information matrix. Furthermore, some important practical issues such as the efficient recovery of the state estimate and the associated covariance matrix need further work.

The major contributions of this book include three new exactly sparse information filters for SLAM: one is achieved by decoupling the localization and mapping processes in SLAM; the other two are aimed at SLAM in large environments through joining many small scale local maps. In the first algorithm, D-SLAM, it is shown that SLAM can be performed in a decoupled manner in which the localization and mapping are two separate yet concurrent processes. This formulation of SLAM results in a new and natural way to achieve the sparse information matrix without any approximation. In the second algorithm, the relative information present in each local map is first extracted, and then used to build a global map based on the D-SLAM framework. Both these algorithms, while computationally efficient, incur some information loss. The third algorithm that modifies the global map state vector by incorporating robot start and end poses of each local map, completely avoids the information loss while maintaining the sparseness of the information matrix and associated computational advantages. These algorithms are developed based on a detailed analysis of the sparse structure and the evolution of the SLAM information matrix.

Two efficient methods for recovering the state estimate and desired columns of the associated covariance matrix from the output of the EIF are also developed. These methods exploit the gradual evolution of the SLAM information matrix, and allow the EIF-based SLAM algorithms presented in this book to be implemented at a computational cost that is approximately linearly proportional to the size of the map.

Apart from above contributions, three key fundamental properties of the SLAM problem and the EKF/EIF SLAM approaches, which include observability, convergence and consistency, are analyzed in detail, and proofs of all results are provided. The basic principles of other major SLAM approaches, including particle filters based and graph based approaches, are also briefly introduced.

This book is based on the PhD thesis of the first author at the University of Technology, Sydney, supervised by the second and third authors. Materials describing the contributions of a number of other researchers have also been added into the book to make it more informative for students and researchers in the field of robotics.

Zhan Wang
Shoudong Huang
Gamini Dissanayake

Sydney, Australia
July, 2009

Acknowledgments

This work is supported by the ARC Centre of Excellence programme, funded by the Australian Research Council (ARC) and the New South Wales State Government.

Contents

Chapter 1

Introduction

Simultaneous Localization and Mapping (SLAM) is the process of building a map of an environment while concurrently generating an estimate for the location of the robot. SLAM provides a mobile robot with the fundamental ability to localize itself and the features in the environment without a prior map, which is essential for many navigation tasks.

Stochastic estimation techniques such as the Extended Kalman Filter (EKF) or the Extended Information Filter (EIF) have been used to solve the SLAM problem. In the traditional EKF SLAM algorithm, the state consisting of the robot pose and all feature locations is estimated using an EKF and all correlations among the states are maintained to achieve an optimal estimate. In EKF/EIF based SLAM formulations, the probabilistic properties of the state estimate is represented by one vector (state vector/information vector) and one square matrix (covariance matrix/information matrix). The covariance matrix is dense and requires a significant computational effort to maintain, particularly for large scale problems. Therefore, if the information matrix (the inverse of the covariance matrix) is sparse, significant savings of computational cost can be achieved.

This book focuses on the computationally efficient SLAM algorithms when range and bearing observations from a robot to point features are available. Three new SLAM algorithms that achieve exact sparseness of the information matrix are presented. In the first algorithm, it is shown that the SLAM problem can be reformulated such that the mapping and localization can be treated as two concurrent yet separated processes, resulting in an exactly sparse information matrix. The second and third algorithms provide two different strategies for joining small local maps to build a global map while maintaining the sparseness of the information matrix for dealing with the large scale SLAM problem. Important aspects of EIF-based SLAM algorithms such as state and covariance recovery, and data association are also addressed.

1.1 The SLAM Problem and Its Applications

1.1.1 *Description of the SLAM Problem*

The setting for the SLAM problem (2D) is that of a robot moving in an environment consisting of a population of features as shown in Figure 1.1. The robot is equipped with proprioceptive sensors that can measure its own motion and exteroceptive sensors that can take measurements of the relative location between the robot and nearby features. The objective of the SLAM problem is to estimate the position and orientation of the robot together with the locations of all the features.

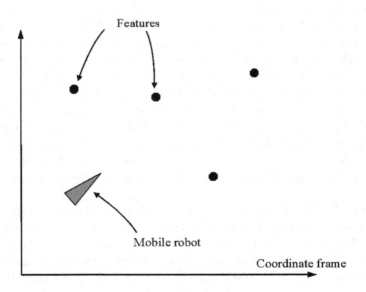

Fig. 1.1 The setting of the SLAM problem.

1.1.2 *Applications of SLAM*

When a mobile robot is expected to operate in an unknown or partially known environment, for example in urban search and rescue or underwater exploration, efficiently acquiring a geometric representation of the environment is crucial. Thus SLAM plays a key role in any successful deployment of mobile robots in such domains. In the past few years, SLAM algorithms have been developed to suit a large number of application domains. In the following some interesting examples are provided.

Risks associated with working in the subsea hostile environment has been the key motivation for the development of Autonomous Underwater Vehicles (AUVs). Lack of external infrastructure such as Global Positioning System (GPS), that is inaccessible underwater, has made SLAM a crucial component of the navigation suites of AUVs used for subsea survey missions and exploration for ocean science. Typically on-board vision or acoustics devices are used to gather information about the environment and to be subsequently used in SLAM algorithms to provide essential geometric information required for navigating the AUV. Figure 1.2 shows the AUV, SIRIUS, from the Australian Centre for Field Robotics, University of Sydney. SIRIUS is equipped with a high resolution stereo imaging system with strobes, a multibeam sonar, depth and conductivity/temperature sensors, Doppler Velocity Log (DVL), Ultra Short Baseline Acoustic Positioning System (USBL) and a forward looking obstacle avoidance sonar.

Fig. 1.2 The SIRIUS (Courtesy of Australian Centre for Field Robotics, University of Sydney).

One of the most important applications of Unmanned Aerial Vehicle (UAV) is aerial observations, which can be used for terrain mapping, environmental surveys, surveillance or disaster response. In these scenarios, either mapping the unknown environment or localizing one or several targets in the environment is necessary. SLAM becomes essential when the position of the UAV is unknown, typically in defense scenarios where GPS signals are considered somewhat unreliable due to the potential for jamming. Figure 1.3 shows the UAV developed in the Australian Centre for Field Robotics, University of Sydney. The UAV uses visual sensors to make observations of the environment.

Fig. 1.3   A UAV in action (Courtesy of Australian Centre for Field Robotics, University of Sydney).

Search and rescue robots provide responders with useful information about the disaster scenes using on-board sensing facilities. They help with determining the victim's condition, providing an accurate victim location and enabling victim recovery. The capabilities of generating easily understandable and accurate maps of the environment and providing specific directions of how to reach the victims from the location of the responders are essential. Figure 1.4 shows the Packbot robot equipped with the FasTac Kit entering a dark area.

1.2   Summary of SLAM Approaches

In the past decade, the SLAM problem has been the subject of extensive research with a number of robotics research groups contributing to

Fig. 1.4 Packbot with FasTac Kit (Courtesy of iRobot).

make substantial progress in this area (see for example [Dissanayake *et al.* (2001); Frese *et al.* (2005); Thrun *et al.* (2004a)] and the references therein). Following the pioneer work on the estimation-theoretic based approach to the SLAM problem by [Smith *et al.* (1987)] and [Moutarlier and Chatila (1989b)], considerable research has been focused on this direction. The body of literature in SLAM is large and an extensive review of the recent work in SLAM has appeared in [Durrant-Whyte and Bailey (2006)] and [Bailey and Durrant-Whyte (2006)]. Details of several state of the art SLAM algorithms can be found in the classic textbook [Thrun *et al.* (2005)]. Classification of major SLAM approaches are provided in [Siciliano and Khatib (2008)].

1.2.1 *EKF/EIF based SLAM Approaches*

Let the states of the robot be denoted by $X_r = [x_r, y_r, \phi_r]^T$, in which (x_r, y_r) represents the position and ϕ_r represents the orientation of the robot. Let the states of features be denoted by $X_m = [x_1, y_1, x_2, y_2, \cdots]^T$, in which (x_j, y_j) represents the position of feature j. Therefore the state vector is $X = [X_r^T, X_m^T]^T$ which contains both the robot states X_r and the feature states X_m.

The kinematic model of the robot which relates the robot poses at time steps k and $k + 1$, $X_r(k)$ and $X_r(k + 1)$, is given by

$$X_r(k + 1) = f(X_r(k), u(k)) + w_r(k) \qquad (1.1)$$

where $u(k)$ indicates the control measurement at time step k, $w_r(k)$ is the process noise assumed to be white Gaussian with zero mean and a covariance Q_r, and the function f depends on the robot model.

The state transition equation for features is

$$X_m(k+1) = X_m(k) \tag{1.2}$$

as features are assumed to be stationary.

Therefore, the state transition model for the whole system may be written as

$$X(k+1) = F(X(k), u(k)) + w(k) \tag{1.3}$$

i.e.

$$\begin{bmatrix} X_r(k+1) \\ X_m(k+1) \end{bmatrix} = \begin{bmatrix} f(X_r(k), u(k)) \\ X_m(k) \end{bmatrix} + \begin{bmatrix} w_r(k) \\ 0 \end{bmatrix} \tag{1.4}$$

where the covariance matrix of $w(k)$ is denoted as Q.

The observation model is

$$z(k) = H(X(k)) + v(k) \tag{1.5}$$

where H defines the nonlinear coordinate transformation from the state to the observation $z(k)$. The observation noise $v(k)$ is assumed to be white Gaussian with zero mean and a covariance R, and uncorrelated with the process noise $w_r(k)$.

For range-bearing SLAM, if the observation $z(k)$ contains the range and bearing measurements from the robot pose, $X_r(k) = [x_r(k), y_r(k), \phi_r(k)]^T$, to one feature, $X_{mi} = [x_{mi}, y_{mi}]^T$, then the observation model is of the form

$$\begin{aligned} r(k) &= \sqrt{(x_{mi} - x_r(k))^2 + (y_{mi} - y_r(k))^2} + v_r(k) \\ \theta(k) &= \arctan(\tfrac{y_{mi}-y_r(k)}{x_{mi}-x_r(k)}) - \phi_r(k) + v_\theta(k). \end{aligned} \tag{1.6}$$

From the probabilistic perspective, the objective of SLAM is to obtain the joint probability distribution of the robot location and feature locations conditioned on the information from the observation and robot motion. With the noise in $w(k)$ and $v(k)$ being assumed to be white Gaussian, EKF can be used to solve SLAM as a nonlinear estimation problem. The mean of the state vector and the associated covariance matrix then are sufficient to describe the joint Gaussian probability distribution.

The estimation-theoretic SLAM algorithm is based on the work of [Smith *et al.* (1990)]. It uses the EKF to estimate X and compute the associated covariance matrix P, i.e.

$$P = \begin{bmatrix} P_{rr} & P_{rm} \\ P_{rm}^T & P_{mm} \end{bmatrix}, \tag{1.7}$$

where P_{rr} is the submatrix corresponding to the robot pose estimate, P_{mm} is the submatrix corresponding to the map estimate and P_{rm} is the correlation matrix.

Suppose at time step k, the estimates of the robot pose and feature locations are $\hat{X}(k|k) = [\hat{X}_r(k|k)^T, \hat{X}_m(k|k)^T]^T$ and the associated covariance matrix is $P(k|k)$. The EKF proceeds recursively in two stages:

(1) Predict the current robot pose and the feature locations using the state transition model of the whole system and compute the state estimate covariance matrix $P(k+1|k)$:

$$\begin{aligned} \hat{X}_r(k+1|k) &= f(\hat{X}_r(k|k), u(k)) \\ \hat{X}_m(k+1|k) &= \hat{X}_m(k|k) \\ P(k+1|k) &= \nabla F_X P(k|k) \nabla F_X^T + Q, \end{aligned} \tag{1.8}$$

where ∇F_X is the Jacobian of F with respect to X evaluated at $\hat{X}(k|k)$.

(2) Update the state estimate and the associated covariance matrix using the observation $z(k+1)$:

$$\begin{aligned} \hat{X}(k+1|k+1) &= \hat{X}(k+1|k) + K(k+1)[z(k+1) - H(\hat{X}(k+1|k))] \\ P(k+1|k+1) &= P(k+1|k) - K(k+1)S(k+1)K(k+1)^T, \end{aligned} \tag{1.9}$$

where

$$\begin{aligned} K(k+1) &= P(k+1|k) \nabla H_X^T S(k+1)^{-1} \\ S(k+1) &= \nabla H_X P(k+1|k) \nabla H_X^T + R, \end{aligned} \tag{1.10}$$

and ∇H_X is the Jacobian of H with respect to X evaluated at $\hat{X}(k+1|k)$.

The mathematical dual of the EKF, the EIF, has also been used to solve the SLAM problem. The EIF uses the inverse covariance matrix, which is termed as the information matrix, and the information vector to define the joint Gaussian probability distribution:

$$\begin{aligned} I &= P^{-1} \\ i &= I\hat{X} \end{aligned} \tag{1.11}$$

where I is the information matrix and i is the information vector.

Suppose at time step k, the information matrix is $I(k|k)$ and the information vector is $i(k|k)$. The EIF proceeds recursively in two stages in the same manner as EKF but the computational effort required is different [Eustice *et al.* (2006)]:

(1) The prediction step involves more computation because the operation of marginalizing out the past robot pose is more complex in the information form:

$$I(k+1|k) = [\nabla F_X I(k|k)^{-1} \nabla F_X^T + Q]^{-1}$$
$$i(k+1|k) = I(k+1|k) F(\hat{X}(k|k), u(k)) \tag{1.12}$$

where ∇F_X is the Jacobian of F with respect to X evaluated at $\hat{X}(k|k)$ [Manyika and Durrant-Whyte (1994)].

(2) The update step incorporates the information in the observation $z(k+1)$ by the addition operation to the previous information matrix and information vector:

$$I(k+1|k+1) = I(k+1|k) + \nabla H_X^T R^{-1} \nabla H_X$$
$$i(k+1|k+1) = i(k+1|k) + \nabla H_X^T R^{-1}[z(k+1)$$
$$- H(\hat{X}(k+1|k)) + \nabla H_X \hat{X}(k+1|k)] \tag{1.13}$$

where ∇H_X is the Jacobian of the function H with respect to all the states evaluated at the current state estimate $\hat{X}(k+1|k)$.

1.2.2 *Other SLAM Approaches*

In this section, the fundamental principles of other major SLAM approaches, including particle filters based and graph based approaches, are generally and briefly introduced, mainly following [Siciliano and Khatib (2008)] and [Durrant-Whyte and Bailey (2006)].

1.2.2.1 *Particle filters based SLAM approaches*

Particle filters perform recursive Monte carlo sampling to estimate probability densities which are represented by particles. Based on the sequential importance sampling, particle filters gained practical popularity after effective resampling steps were introduced in [Salmond and Gordon (2000)]. In recent years, the availability of increased computing power has made particle filters an effective practical proposition for many estimation tasks.

Particle filters have been used to estimate the full SLAM posterior. In a general full SLAM problem, suppose there are $l+1$ robot poses,

$X_{r0}, \ldots, X_{rl}$, and s features, $X_{m1}, \ldots, X_{ms}$, in the state vector. Let the control measurement and observation of each time step be u_k and $z_k (k = 1 \ldots l)$ respectively.[1] The full SLAM posterior is

$$p(X_{r(0:l)}, X_{m(1:s)} | z_{1:l}, u_{1:l}). \tag{1.14}$$

However, the direct use of particle filters to perform this estimation is infeasible, as the space of states including features and all robot poses is huge. Computationally, particle filters scale exponentially with the dimension of the state space. In FastSLAM [Montemerlo *et al.* (2002)], Rao-Blackwellization and conditional independence have been exploited to conquer the computational burden.

The sampling space can be reduced by Rao-Blackwellization. Suppose a probability density function $p(a, b)$ is to be computed, where a and b are random variables. Instead of drawing samples from the joint distribution, one only need to draw samples from $p(a)$, if $p(b|a)$ can be represented analytically. This can be seen clearly from the product rule, $p(a, b) = p(a)p(b|a)$. It turns out that sampling from the reduced space produces more accurate representation of the joint distribution. The full SLAM posterior can be written as

$$\begin{aligned} &p(X_{r(0:l)}, X_{m(1:s)} | z_{1:l}, u_{1:l}) \\ &= p(X_{r(0:l)} | z_{1:l}, u_{1:l}) p(X_{m(1:s)} | X_{r(0:l)}, z_{1:l}). \end{aligned} \tag{1.15}$$

As in FastSLAM, $p(a)$ corresponds to $p(X_{r(0:l)} | z_{1:l}, u_{1:l})$ and $p(b|a)$ corresponds to $p(X_{m(1:s)} | X_{r(0:l)}, z_{1:l})$. By applying Rao-Blackwellization, it is only necessary to sample the state space that represents the robot full trajectory. In other words, the robot full trajectory is represented by samples, and a map is attached to each sample in close-form.

Conditional independence of the features in the map, provided the full robot trajectory, is exploited to efficiently represent and estimate the map. Given the full robot trajectory, the map can be decomposed as

$$p(X_{m(1:s)} | X_{r(0:l)}, z_{1:l}) = \prod_i p(X_{mi} | X_{r(0:l)}, z_{1:l}). \tag{1.16}$$

FastSLAM uses the EKF to estimate the map attached to each particle. Decoupling the high-dimensional probability density function of the whole map results in a representation by many low-dimensional probability den-

---

[1]Notations in different sections of this book may have different meaning according to what is described in the context.

sity functions, each of which corresponds to one feature. Therefore, the computation of estimating the map by the EKF is significantly reduced.

The general Rao-Blackwellized particle filters for SLAM can be summarized in four steps [Durrant-Whyte and Bailey (2006)]. Suppose at the time step $k - 1$, the full SLAM posterior is represented by $\{w_{k-1}^{(i)}, X_{r(0:k-1)}^{(i)}, p(X_m | X_{r(0:k-1)}^{(i)}, z_{1:k-1})\}$, where w is the importance weight assigned to each particle and the superscript indicates the particle index.

(1) For each particle, draw a sample $X_{rk}^{(i)}$ from a computed proposal distribution. The new sample is then added to the particle history.

(2) The importance weight of each particle is updated according to the importance function, based on the observation and process models as well as the proposal distribution. The importance weights of all particles need to be normalized so that they sum to 1.

(3) Resampling is performed when necessary. From the existing particles, a new set of particles are drawn with replacement. The probability of drawing from the existing particles is proportional to the importance weight, as the particles that fit the measurements better should have greater chances for survival. Then the selected particles are assigned uniform weight.

(4) For each particle, the attached map is updated using observations to features from the current robot pose by an EKF.

1.2.2.2   *Graph based SLAM approaches*

Graph based methods have also been developed to address the full SLAM problem. The feature locations and robot poses can be represented as nodes of a graph, while the control measurements and observations constitute the edges between the relevant nodes. As each robot node is only connected to the feature nodes where observations exist and there are no connections among feature nodes, the graph is sparse. Nonlinear optimization techniques that operate on this sparse graph are used to find the optimal values for feature locations and robot poses which maximize the likelihood of the control measurements and observations.

Consider the full SLAM problem as described in Section 1.2.2.1. There are $l + 1$ robot poses, $X_{r0}, \ldots, X_{rl}$, and s features, $X_{m1}, \ldots, X_{ms}$, in the state vector. The control measurement and observation of each time step are u_k and $z_k(k = 1 \ldots l)$ respectively, and the SLAM posterior is

$$p(X_{r(0:l)}, X_{m(1:s)} | z_{1:l}, u_{1:l}). \tag{1.17}$$

Assuming the observations are conditionally independent, the whole posterior can be decomposed as [Thrun *et al.* (2005)]

$$c\, p(X_{r0})\prod_k\left[p(X_{rk}|X_{r(k-1)},u_k)\prod_i p(z_k^i|X_{rk},X_{mj_i})\right] \qquad (1.18)$$

where c is the normalizer and $p(X_{r0})$ is the prior. Note z_k^i is the i-th observation in z_k and with assumed data association X_{mj_i} is the feature involved in this observation. The terms $p(X_{rk}|X_{r(k-1)},u_k)$ and $p(z_k^i|X_{rk},X_{mj_i})$ are actually the probabilistic density function form of state transition model for robot pose and the observation model as in equations (1.3) and (1.5) respectively.

As the noises are assumed to be Gaussian, the state transition model can be written as

$$
\begin{aligned}
&p(X_{rk}|X_{r(k-1)},u_k)\\
&\propto \exp\{-\tfrac{1}{2}[X_{rk}-f(X_{r(k-1)},u_k)]^T Q_r^{-1}[X_{rk}-f(X_{r(k-1)},u_k)]\}
\end{aligned}
\qquad (1.19)
$$

and the observation model can be written as

$$
\begin{aligned}
&p(z_k|X_{rk},X_{m(1:s)})\\
&\propto \exp\{-\tfrac{1}{2}[z_k-H(X_{rk},X_{m(1:s)})]^T R^{-1}[z_k-H(X_{rk},X_{m(1:s)})]\},
\end{aligned}
\qquad (1.20)
$$

where Q_r and R are corresponding covariance matrices for process and observation noises. Then the negative log posterior can be written as

$$
\begin{aligned}
&-log\, p(X_{r(0:l)},X_{m(1:s)}|z_{1:l},u_{1:l})\\
&\propto \sum_k[X_{rk}-f(X_{r(k-1)},u_k)]^T Q_r^{-1}[X_{rk}-f(X_{r(k-1)},u_k)]\\
&\quad +\sum_k[z_k-H(X_{rk},X_{m(1:s)})]^T R^{-1}[z_k-H(X_{rk},X_{m(1:s)})]\\
&= \Psi(X)
\end{aligned}
\qquad (1.21)
$$

where $X = [X_r^T, X_m^T]^T$.

Maximizing the likelihood of all control measurements and observations requires finding the optimal value for the following cost function

$$\hat{X} = \underset{X}{\operatorname{argmin}}\Psi(X). \qquad (1.22)$$

Optimization techniques such as gradient descent can be used. The data association problem can be easily considered in the optimization, as the knowledge of data association can be readily incorporated into the cost function in the form of additional constraints.

1.3   Key Properties of SLAM

In this section, three important properties of the SLAM problem and the EKF/EIF SLAM approaches are examined in detail: observability, convergence and consistency.

1.3.1   *Observability*

Before discussing the observability property, it is important to recognize that there are two versions of the SLAM problem. Absolute SLAM or world-centric SLAM attempts to estimate the absolute global location of a mobile robot and features present in the environment, using observations (range and/or bearing) and control measurements made from the mobile robot. Most work available in literature on SLAM observability addresses the absolute SLAM problem. As observations which are available in a totally unknown environment only relate the robot location to the features present in this environment and control measurements only relate the two relevant robot locations, it is easy to see that the initial location of the robot with respect to a world centric global reference frame cannot be obtained using these data, making the absolute version of SLAM unobservable. In relative SLAM, the robot location and the map are computed relative to a coordinate frame attached to the initial robot pose. This is equivalent to assuming perfect knowledge of the initial robot location and is almost universally used in practical demonstrations of SLAM.

From the control theoretic point of view, observability is the ability to uniquely compute the system initial state from a sequence of control actions and observations. Results from a number of publications that use techniques from control theories demonstrate that, as expected, the absolute version of SLAM is not observable [Lee *et al.* (2006)]. Therefore, at least some observations to features whose locations are known with respect to the global reference frame, or input from an external sensor, such as GPS, indicating the location of the robot in the global reference frame are required to solve the world-centric SLAM problem. In [Vidal-Calleja *et al.* (2007)], SLAM equations are approximated by a piecewise constant system and world-centric SLAM with one known feature and two time segments is said to be observable. It is argued in [Lee *et al.* (2006)] that using linearised models within control theoretic observability analysis could lead to incorrect conclusions and they present a non-linear analysis of SLAM observability. It is shown that, in contrast to results present in the literature, the world-centric SLAM problem with one known feature is unobservable. These authors also demonstrate, analytically as well as through computer simulations, that world-centric SLAM with a range-bearing sensor is observable only when absolute locations of two features in the environment are available. In case of relative SLAM, the appropriate question on observability is whether the system initial state can be uniquely determined if a part of the initial state, i.e. the initial location of the robot, is available. This question can not be answered by applying the formulae available from control theory without modifying the system equations for SLAM to include the initial pose of the robot.

Alternatively, from the estimation theoretic point of view, observability condition is evaluated by examining the Fisher information matrix (FIM) [Bar-Shalom *et al.* (2001)]. A linearised version of the SLAM problem is used to derive the FIM for world-centric SLAM in [Andrade-Cetto and Sanfeliu (2005)]. The fact that the FIM is singular is also demonstrated. It is also shown that if the location of one feature is known, SLAM problem with a Monobot (one-dimensional version where a robot moves on a straight line and observes features that are located on this line) is observable. This analysis is extended to a two-dimensional planar robot using a linearised model to show that world-centric SLAM is observable if the global location of one feature is available. This conclusion contradicts the results from [Lee *et al.* (2006)] and the intuitive notion that two absolute locations are needed to anchor a planar map that has three degrees of freedom. This is perhaps an artifact of the linearization process.

The observability of the relative SLAM problem is of most interest, as this is the SLAM problem which is addressed by the majority of the SLAM literature. In particular, the range-bearing relative SLAM problem is discussed in this book. As such, in this section, the observability of this SLAM scenario is analyzed by deriving the FIM for the SLAM problem taking consideration of all the robot motion and observation information. The observability of other SLAM scenarios (such as world-centric range and bearing SLAM) can also be analyzed by the developed method, and the readers are referred to [Wang and Dissanayake (2008)].

1.3.1.1 *Fisher Information Matrix of the SLAM problem*

In contrast to traditional EKF based SLAM algorithms, consider a state vector that consists of all robot poses and all the features in the environment. Let the states of the j-th robot pose be denoted by $X_{rj} = [x_{rj}, y_{rj}, \phi_{rj}]^T$ and the states of the i-th feature be denoted by $X_{mi} = [x_{mi}, y_{mi}]^T$. The whole state vector is, therefore, $X = [X_{r1}^T, X_{r2}^T, \ldots, X_{m1}^T, X_{m2}^T, \ldots]^T$.

Process model

In order to facilitate the analysis of observability, the information contained in the control signals is captured using the relationship between two successive robot poses. As shown in Figure 1.5, $D(k) = [x_{rel}(k), y_{rel}(k), \phi_{rel}(k)]^T$ is the relative position of pose p_{k+1} with respect to the origin in the coordinate frame attached to the robot at pose p_k [Frese (2004)].

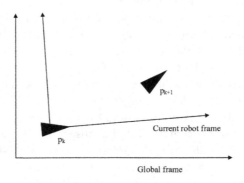

Fig. 1.5   Relative position.

The relative position $D(k)$ can be expressed as

$$D(k) = f(X_r(k), X_r(k+1)) + w(k) \tag{1.23}$$

where $X_r(k)$ and $X_r(k+1)$ are the robot poses at the time step k and $(k+1)$ respectively, and $w(k)$ is the process noise assumed to be white Gaussian with zero mean and a covariance $Q(k)$.

This equation can be expanded to result in

$$\begin{bmatrix} x_{rel}(k) \\ y_{rel}(k) \\ \phi_{rel}(k) \end{bmatrix} = \begin{bmatrix} f_1(X_r(k), X_r(k+1)) \\ f_2(X_r(k), X_r(k+1)) \\ f_3(X_r(k), X_r(k+1)) \end{bmatrix} + w(k) \tag{1.24}$$

where

$$f_1(X_r(k), X_r(k+1)) = (x_r(k+1) - x_r(k))\cos(\phi_r(k))$$
$$+ (y_r(k+1) - y_r(k))\sin(\phi_r(k)) \tag{1.25}$$

$$f_2(X_r(k), X_r(k+1)) = (y_r(k+1) - y_r(k))\cos(\phi_r(k))$$
$$- (x_r(k+1) - x_r(k))\sin(\phi_r(k)) \tag{1.26}$$

$$f_3(X_r(k), X_r(k+1)) = \phi_r(k+1) - \phi_r(k). \tag{1.27}$$

The relative position $D(k)$ is related to the robot forward and angular velocities through an appropriate non-linear kinematic model. It is possible to combine the measurements of the control signals available between the two robot poses p_k and p_{k+1}, to obtain the relative position $D(k)$. Assuming that the control signal measurements are corrupted with zero mean white Gaussian noise, the covariance matrix $Q(k)$ of $D(k)$ can also be easily obtained through an estimation process, for example, using an EKF.

Clearly $D(k)$ can be expressed as a function of the states and it contains the information obtained from a sensor that measures the control signals. Thus using $D(k)$ as an observation in the analysis of observability allows the effect of the information content present in the measured control signals to be explicitly captured. It is also important to note that knowledge of one robot pose, implies some knowledge of all robot poses as these entities are directly related through the robot kinematic model and measured control signals. Therefore, in the context of observability analysis, the reformulated SLAM problem behaves in a manner identical to the traditional SLAM formulation.

Observation model

The observation model is

$$Z(k) = H(X(k)) + \upsilon(k) \tag{1.28}$$

where the state vector X contains both the robot pose states and the feature states, H defines the nonlinear coordinate transformation from the state to the observation $Z(k)$. The observation noise $\upsilon(k)$ is assumed to be white Gaussian with zero mean and a covariance R.

For range-bearing SLAM, if the observation $Z(k)$ contains the range and bearing measurements from one robot pose, $X_r = [x_r, y_r, \phi_r]^T$, to one feature, $X_m = [x_m, y_m]^T$, the observation model is of the form

$$r(k) = \sqrt{(x_m - x_r)^2 + (y_m - y_r)^2} + \upsilon_r(k)$$

$$\theta(k) = \arctan(\tfrac{y_m - y_r}{x_m - x_r}) - \phi_r + \upsilon_\theta(k). \tag{1.29}$$

Fisher Information Matrix

In a general form, let all the available observations, $D_i, (i = 1 \ldots m)$ and $Z_j, (j = 1 \ldots n)$, be combined to form a vector M

$$M = \begin{bmatrix} D_1^T & \ldots & D_m^T & Z_1^T & \ldots & Z_n^T \end{bmatrix}^T \tag{1.30}$$

which has the dimension $(3m + 2n) \times 1$.

The observation equation can now be written as

$$M = g(X) + \nu \tag{1.31}$$

where ν is the observation noise which is Gaussian with zero mean and a covariance W. W can be easily constructed using Q_i, the covariance matrix of $D_i (i = 1 \ldots m)$, and R_j, the covariance matrix of $Z_j (j = 1 \ldots n)$.

For nonrandom vector parameter estimation, the FIM for an unbiased estimator is defined as [Bar-Shalom *et al.* (2001)]

$$J \triangleq E\{[\nabla_X ln\Lambda(X)][\nabla_X ln\Lambda(X)]^T\} \tag{1.32}$$

where $\Lambda(X) \triangleq p(M|X)$ is the likelihood function. Note that the partial derivatives should be calculated at the true value of X.

From equation (1.31),

$$\begin{aligned} \Lambda(X) &\triangleq p(M|X) \\ &= N(M; g(X), W) \\ &= ce^{-1/2[M-g(X)]^T W^{-1}[M-g(X)]} \end{aligned} \tag{1.33}$$

where $N(.)$ indicates Gaussian distribution and c is some constant.

Then

$$\begin{aligned} &\nabla_X ln\Lambda(X) \\ &= \nabla_X\{-1/2[M - g(X)]^T W^{-1}[M - g(X)]\} \\ &= J_G^T W^{-1}[M - g(X)] \end{aligned} \tag{1.34}$$

where J_G is Jacobian of function g with respect to X, evaluated at the true value of X.

Note that the expectation

$$\begin{aligned} &E\{[M - g(X)][M - g(X)]^T\} \\ &= \int_{-\infty}^{+\infty}[M - g(X)][M - g(X)]^T p(M|X)dM \\ &= W. \end{aligned} \tag{1.35}$$

By using (1.34) and (1.35), the FIM can be derived as

$$
\begin{aligned}
J &\triangleq E\{[\nabla_X ln\Lambda(X)][\nabla_X ln\Lambda(X)]^T\} \\
&= E\{J_G^T W^{-1}[M - g(X)][M - g(X)]^T W^{-1} J_G\} \\
&= J_G^T W^{-1} J_G.
\end{aligned} \tag{1.36}
$$

When true values are not available, Jacobians should be evaluated at the current estimates [Bar-Shalom *et al.* (2001)].

1.3.1.2 *Observability analysis of the SLAM problem*

If the FIM in equation (1.36) is singular, then the system is unobservable [Bar-Shalom *et al.* (2001)]. Examining the structure of J_G and W in equation (1.36) enables the observability of the system and that of each state to be analyzed.

The covariance matrix W is of the form

$$
W = diag(Q_1, Q_2, \ldots, R_1, R_2, \ldots). \tag{1.37}
$$

It is easily seen that W^{-1} is also symmetric and positive definite. Thus there must exist a lower triangular matrix L with positive diagonal elements that fulfills

$$
W^{-1} = LL^T. \tag{1.38}
$$

Note that L is nonsingular.

The FIM can now be written as

$$
\begin{aligned}
J &= J_G^T LL^T J_G \\
&= (L^T J_G)^T (L^T J_G).
\end{aligned} \tag{1.39}
$$

Due to the fact that L is not singular,

$$
rank(J) = rank(L^T J_G) = rank(J_G). \tag{1.40}
$$

The Jacobian J_G is of dimension $a \times b$, where a is the total number of observations and b is the dimension of the state vector (also the dimension of J). When $a < b$, FIM is obviously not full rank. However, $a \geq b$ is what typically occurs. In this case, the sufficient and necessary condition for FIM to be nonsingular (full rank) is that the Jacobian J_G is full column rank.

In the scenarios where the FIM is not full rank, it is particularly interesting to examine the reasons why the Jacobian J_G is not full column rank. This can lead to the determination of the conditions required to make the system observable.

Consider the scenario of $2D$ range-bearing relative SLAM with one unknown feature, which is observed from two distinct robot poses. The state vector is $X = [X_{r1}^T, X_{r2}^T, X_{m1}^T]^T$. The observation vector is $M = [D_0^T, D_1^T, Z_1^T, Z_2^T]^T$, in which D_0 provides the relationship between the origin and X_{r1}, D_1 relates X_{r1} and X_{r2}, Z_1 and Z_2 are range and bearing observations taken at X_{r1} and X_{r2} respectively.

The observation Jacobian J_G for this problem is

$$
\begin{bmatrix}
1 & 0 & 0 & 0 & 0 & 0 & 0 & 0 \\
0 & 1 & 0 & 0 & 0 & 0 & 0 & 0 \\
0 & 0 & 1 & 0 & 0 & 0 & 0 & 0 \\
\star & \star & \star & \cos(\phi_{r1}) & \sin(\phi_{r1}) & 0 & 0 & 0 \\
\star & \star & \star & -\sin(\phi_{r1}) & \cos(\phi_{r1}) & 0 & 0 & 0 \\
0 & 0 & -1 & 0 & 0 & 1 & 0 & 0 \\
\star & \star & 0 & 0 & 0 & 0 & \frac{\partial g_7}{x_{m1}} & \frac{\partial g_7}{y_{m1}} \\
\star & \star & -1 & 0 & 0 & 0 & \frac{\partial g_8}{x_{m1}} & \frac{\partial g_8}{y_{m1}} \\
0 & 0 & 0 & \star & \star & 0 & \frac{\partial g_9}{x_{m1}} & \frac{\partial g_9}{y_{m1}} \\
0 & 0 & 0 & \star & \star & -1 & \frac{\partial g_{10}}{x_{m1}} & \frac{\partial g_{10}}{y_{m1}}
\end{bmatrix}
\tag{1.41}
$$

where $\star$ represents some irrelevant value. The expressions g_7, g_8, g_9, g_{10} take the form in the observation equations in (1.29). In determining the rank of J_G in (1.41), it is only interesting to examine if the last two columns of J_G are dependent. From (1.29), the elements in these two columns can be expanded as

$$
\frac{\partial g_7}{\partial x_{m1}} = \frac{-(x_{r1} - x_{m1})}{\sqrt{(x_{r1} - x_{m1})^2 + (y_{r1} - y_{m1})^2}}
\tag{1.42}
$$

$$
\frac{\partial g_7}{\partial y_{m1}} = \frac{-(y_{r1} - y_{m1})}{\sqrt{(x_{r1} - x_{m1})^2 + (y_{r1} - y_{m1})^2}}
\tag{1.43}
$$

$$
\frac{\partial g_8}{\partial x_{m1}} = \frac{(y_{r1} - y_{m1})}{(x_{r1} - x_{m1})^2 + (y_{r1} - y_{m1})^2}
\tag{1.44}
$$

$$
\frac{\partial g_8}{\partial y_{m1}} = \frac{-(x_{r1} - x_{m1})}{(x_{r1} - x_{m1})^2 + (y_{r1} - y_{m1})^2}
\tag{1.45}
$$

$$\frac{\partial g_9}{\partial x_{m1}} = \frac{-(x_{r2} - x_{m1})}{\sqrt{(x_{r2} - x_{m1})^2 + (y_{r2} - y_{m1})^2}} \tag{1.46}$$

$$\frac{\partial g_9}{\partial y_{m1}} = \frac{-(y_{r2} - y_{m1})}{\sqrt{(x_{r2} - x_{m1})^2 + (y_{r2} - y_{m1})^2}} \tag{1.47}$$

$$\frac{\partial g_{10}}{\partial x_{m1}} = \frac{(y_{r2} - y_{m1})}{(x_{r2} - x_{m1})^2 + (y_{r2} - y_{m1})^2} \tag{1.48}$$

$$\frac{\partial g_{10}}{\partial y_{m1}} = \frac{-(x_{r2} - x_{m1})}{(x_{r2} - x_{m1})^2 + (y_{r2} - y_{m1})^2}. \tag{1.49}$$

It is easy to see that there is no real number α that can satisfy

$$\begin{bmatrix} \frac{\partial g_7}{\partial x_{m1}} \\ \frac{\partial g_8}{\partial x_{m1}} \\ \frac{\partial g_9}{\partial x_{m1}} \\ \frac{\partial g_{10}}{\partial x_{m1}} \end{bmatrix} = \alpha \begin{bmatrix} \frac{\partial g_7}{\partial y_{m1}} \\ \frac{\partial g_8}{\partial y_{m1}} \\ \frac{\partial g_9}{\partial y_{m1}} \\ \frac{\partial g_{10}}{\partial y_{m1}} \end{bmatrix} \tag{1.50}$$

unless $X_{r2} = X_{r1}$. Thus J_G is full column rank and then the FIM is full rank. This result can also be extended to any number of features and observations. Therefore, the range-bearing relative SLAM is observable.

1.3.2 *EKF SLAM Convergence*

This section discusses the convergence properties of EKF based SLAM algorithms. While there have been numerous implementations of various SLAM algorithms, only very few analytical results on the convergence and essential properties of the SLAM algorithms are available. Dissanayake *et al.* provided convergence properties of EKF SLAM and the lower bound on the position uncertainty [Dissanayake *et al.* (2001)]. These results were extended to multi-robot SLAM in [Fenwick *et al.* (2002)]. In [Kim (2004)] some further analysis on the asymptotic behavior for the one dimensional SLAM problem were provided. However, the proofs presented in these papers only deal with simple linear formulations of the SLAM problem.

Almost all practical SLAM implementations need to deal with nonlinear process and observation models. The results in [Dissanayake *et al.* (2001)] are intuitive and many early experiments and computer simulations appear to confirm that the properties of the linear solution can be extended to practical nonlinear problems. In the recent work [Huang and Dissanayake (2007)], proofs of convergence are provided for the nonlinear two-dimensional SLAM problem. It is shown that most of the convergence properties in [Dissanayake *et al.* (2001)] are still true for the nonlinear case provided that the Jacobians used in the EKF equations are evaluated at the true values. It is also shown that the robot orientation uncertainty at the instant when features are first observed has a significant effect on the limit and/or the lower bound of the uncertainties of the feature position estimates.

This section provides both the key convergence properties and the explicit formulas for the covariance matrices of some basic scenarios in the nonlinear two-dimensional EKF SLAM problem with point features observed using a range-and-bearing sensor.

1.3.2.1 *Convergence properties for linear cases*

It is interesting to first examine the convergence property in the linear case. Suppose the process model and observation model are linear, and Kalman Filter is used for the estimation. The state covariance matrix may be written in block form as

$$P = \begin{bmatrix} P_{rr} & P_{rm} \\ P_{rm}^T & P_{mm} \end{bmatrix} \tag{1.51}$$

where P_{rr} is the covariance matrix associated with the robot state estimates; P_{mm} is the covariance matrix associated with the feature state estimates; and P_{rm} is the correlation matrix.

The following three properties are given in [Dissanayake *et al.* (2001)] and the proofs are provided in the linear case.

Theorem 1.1. *The determinant of any submatrix of the map covariance matrix decreases monotonically as successive observations are made.*

Theorem 1.2. *In the limit the feature estimates become fully correlated.*

Theorem 1.3. *In the limit, the lower bound on the covariance matrix associated with any single feature estimate is determined only by the initial covariance in the vehicle estimate P_{0v} at the time of the first sighting of the first feature.*

However, real EKF SLAM deals with nonlinear process and observation models. In the following, the EKF SLAM algorithm is first restated in a form more suitable for theoretical analysis. Then some key convergence properties are proved for the nonlinear case.

1.3.2.2 *Restatement of the EKF SLAM algorithm*

In this section, the EKF SLAM algorithm is restated using slightly different notations and formulas in order to clearly state and prove the results.

State vector in 2D EKF SLAM

The state vector is denoted as[2]

$$X = (\phi, X_r, X_{m1}, \cdots, X_{mN}), \tag{1.52}$$

where ϕ is the robot orientation, $X_r = (x_r, y_r)$ is the robot position, $X_{m1} = (x_1, y_1), \cdots, X_{mN} = (x_N, y_N)$ are the positions of the N point features. Note that the robot orientation ϕ is separated from the robot position because it plays a crucial role in the convergence and consistency analysis.

Process model

The robot process model considered in this section is

$$\begin{bmatrix} \phi(k+1) \\ x_r(k+1) \\ y_r(k+1) \end{bmatrix} = \begin{bmatrix} \phi(k) + f_\phi(\gamma(k), v(k), \delta\gamma, \delta v) \\ x_r(k) + (v(k) + \delta v)T\cos[\phi(k)] \\ y_r(k) + (v(k) + \delta v)T\sin[\phi(k)] \end{bmatrix},$$

and is denoted as

$$\begin{bmatrix} \phi(k+1) \\ X_r(k+1) \end{bmatrix} = f(\phi(k), X_r(k), \gamma(k), v(k), \delta\gamma, \delta v), \tag{1.53}$$

where v, γ are the 'controls', $\delta v, \delta\gamma$ are zero-mean Gaussian noise on v, γ. T is the time interval of one movement step. The explicit formula of function f_ϕ depends on the particular robot. Two examples of this general model are given below.

Example 1. A simple discrete-time robot motion model

$$\begin{aligned} \phi(k+1) &= \phi(k) + (\gamma(k) + \delta\gamma)T \\ x_r(k+1) &= x_r(k) + (v(k) + \delta v)T\cos[\phi(k)] \\ y_r(k+1) &= y_r(k) + (v(k) + \delta v)T\sin[\phi(k)] \end{aligned} \tag{1.54}$$

---

[2]To simplify the notation, the vector transpose operator is omitted. For example, $X, X_r, X_{m1}, \cdots, X_{mN}$ are all column vectors and the rigorous notation should be $X = (\phi, X_r^T, X_{m1}^T, \cdots, X_{mN}^T)^T$.

which can be obtained from a direct discretization of the uni-cycle model
(e.g. [Martinell *et al.* (2005)])

$$\dot{\phi} = \gamma$$
$$\dot{x}_r = v \cos \phi$$
$$\dot{y}_r = v \sin \phi$$
(1.55)

where v is the velocity and γ is the turning rate.

Example 2. A car-like vehicle model (e.g. [Dissanayake *et al.* (2001)])

$$\phi(k+1) = \phi(k) + \frac{(v(k)+\delta v)T \tan(\gamma(k)+\delta\gamma)}{L}$$
$$x_r(k+1) = x_r(k) + (v(k) + \delta v)T \cos[\phi(k)]$$
$$y_r(k+1) = y_r(k) + (v(k) + \delta v)T \sin[\phi(k)]$$
(1.56)

where v is the velocity, γ is the steering angle, and L is the wheel-base of
the vehicle.

The process model of features (assumed stationary) is

$$X_{mi}(k+1) = X_{mi}(k), \quad i = 1, \cdots, N.$$
(1.57)

Thus the process model of the whole system is

$$X(k+1) = F(X(k), \gamma(k), v(k), \delta\gamma, \delta v),$$
(1.58)

where F is the function combining (1.53) and (1.57).

Prediction

Suppose at time k, after the update, the estimate of the state vector is

$$\hat{X}(k|k) = (\hat{\phi}(k), \hat{X}_r(k), \hat{X}_{m1}, \cdots, \hat{X}_{mN}),$$

and the covariance matrix of the estimation error is $P(k|k)$. The prediction
step is given by

$$\hat{X}(k+1|k) = F(\hat{X}(k|k), \gamma(k), v(k), 0, 0),$$
$$P(k+1|k) = \nabla F_{\phi X_r X} P(k|k) \nabla F_{\phi X_r X}^T + \nabla F_{\gamma v} \Sigma \nabla F_{\gamma v}^T,$$
(1.59)

where Σ is the covariance of the control noise $(\delta\gamma, \delta v)$, and $\nabla F_{\phi X_r X}, \nabla F_{\gamma v}$
are given by[3]

$$\nabla F_{\phi X_r X} = \begin{bmatrix} \nabla f_{\phi X_r} & 0 \\ 0 & E \end{bmatrix}, \nabla F_{\gamma v} = \begin{bmatrix} \nabla f_{\gamma v} \\ 0 \end{bmatrix}.$$
(1.60)

Here $\nabla f_{\phi X_r}$ and $\nabla f_{\gamma v}$ are Jacobians of f in (1.53) with respect to the robot
pose (ϕ, X_r) and the control noise $(\delta\gamma, \delta v)$, respectively, evaluated at the
current estimate $\hat{X}(k|k)$.

---

[3] In this section, E and 0 always denote the identity matrix and a zero matrix with an
appropriate dimension, respectively.

For the system described by equation (1.53), the Jacobian with respect to the robot pose is

$$\nabla f_{\phi X_r} = \begin{bmatrix} 1 & 0 & 0 \\ -vT\sin\phi & 1 & 0 \\ vT\cos\phi & 0 & 1 \end{bmatrix}. \qquad (1.61)$$

The Jacobian with respect to the controls, $\nabla f_{\gamma v}$, depends on the detailed formula of function f_ϕ in (1.53). For Example 1,

$$\nabla f_{\gamma v} = \begin{bmatrix} T & 0 \\ 0 & T\cos\phi \\ 0 & T\sin\phi \end{bmatrix}. \qquad (1.62)$$

Measurement model

At time $k+1$, the measurement of i-th feature, obtained using sensor on board the robot, is given by range r_i and bearing θ_i,

$$\begin{aligned} r_i &= \sqrt{(y_i - y_r(k+1))^2 + (x_i - x_r(k+1))^2} + w_{r_i} \\ \theta_i &= \arctan\left(\frac{y_i - y_r(k+1)}{x_i - x_r(k+1)}\right) - \phi(k+1) + w_{\theta_i} \end{aligned} \qquad (1.63)$$

where w_{r_i} and w_{θ_i} are the noises on the measurements.

The observation model can be written in the general form

$$z_i(k+1) = \begin{bmatrix} r_i \\ \theta_i \end{bmatrix} = H_i(X(k+1)) + w_{r_i\theta_i}. \qquad (1.64)$$

The noise $w_{r_i\theta_i}$ is assumed to be Gaussian with zero-mean and covariance matrix $R_{r_i\theta_i}$.

Update

Equation to update the covariance matrix can be written in the information form ([Thrun *et al.* (2005)]) as

$$\begin{aligned} I(k+1|k) &= P(k+1|k)^{-1}, \\ I(k+1|k+1) &= I(k+1|k) + I_{new}, \\ P(k+1|k+1) &= I(k+1|k+1)^{-1}, \end{aligned} \qquad (1.65)$$

where $I(\cdot)$ is the information matrix, I_{new} is the new information obtained from the observation by

$$I_{new} = \nabla H_i^T R_{r_i\theta_i}^{-1} \nabla H_i \qquad (1.66)$$

and ∇H_i is the Jacobian of function H_i evaluated at the current estimate $\hat{X}(k+1|k)$.

The estimate of the state vector can now be updated using

$$\hat{X}(k+1|k+1) = \hat{X}(k+1|k) + W(k+1)\mu(k+1) \qquad (1.67)$$

where

$$\mu(k + 1) = z_i(k + 1) - H_i(\hat{X}(k + 1|k))$$
$$W(k + 1) = P(k + 1|k)\nabla H_i^T S^{-1}(k + 1) \qquad (1.68)$$

and

$$S(k + 1) = R_{r_i\theta_i} + \nabla H_i P(k + 1|k)\nabla H_i^T. \qquad (1.69)$$

Remark 1.1. Using (1.65), (1.66) and the matrix inversion lemma (see equation (A.5) in Appendix A.1),

$$P(k + 1|k + 1) = P(k + 1|k) - P(k + 1|k)\nabla H_i^T$$
$$\cdot (R_{r_i\theta_i} + \nabla H_i P(k + 1|k)\nabla H_i^T)^{-1}\nabla H_i P(k + 1|k), \qquad (1.70)$$

which is the typical EKF update formula.

The Jacobian of the measurement function H_i is

$$\nabla H_i = \begin{bmatrix} 0 & -\frac{dx}{r} & -\frac{dy}{r} & \frac{dx}{r} & \frac{dy}{r} \\ -1 & \frac{dy}{r^2} & -\frac{dx}{r^2} & -\frac{dy}{r^2} & \frac{dx}{r^2} \end{bmatrix} \qquad (1.71)$$

where

$$dx = x_i - x_r(k + 1)$$
$$dy = y_i - y_r(k + 1) \qquad (1.72)$$
$$r = \sqrt{dx^2 + dy^2}.$$

Note that, in the above, all the columns corresponding to features that are not currently being observed have been ignored.

1.3.2.3 *Convergence properties of EKF SLAM*

This section proves some convergence results for 2D nonlinear EKF SLAM. Some of the lengthy proofs are given in Appendix A.2.

The first result is the *monotonically decreasing property* which is the same as Theorem 1.1 (Theorem 1 in [Dissanayake *et al.* (2001)]).

Theorem 1.4. *The determinant of any submatrix of the map covariance matrix decreases monotonically as successive observations are made.*

Proof: This result can be proved in a similar way to that of Theorem 1 in [Dissanayake *et al.* (2001)]. The only difference is that the Jacobians instead of the state transition matrix and observation matrix will be used in the proof. The key point of the proof is "*In the prediction step, the* *covariance matrix of the map does not change; in the update step, the whole covariance matrix is non-increasing".* The details of the proof are omitted.

For 2D nonlinear EKF SLAM, general expressions for the covariance matrix evolution can not be obtained. Therefore, two basic scenarios are considered in the following: (1) the robot is stationary and observes new features many times, and (2) the robot then moves but only observes the same features.

Suppose the robot starts at point **A**, the initial uncertainty of the robot pose is expressed by the covariance matrix

$$P_0 = \begin{bmatrix} p_\phi & p_{xy\phi}^T \\ p_{xy\phi} & P_{xy} \end{bmatrix} \tag{1.73}$$

where p_ϕ is a scalar and P_{xy} is a 2×2 matrix.

The initial information matrix is denoted as

$$I_0 = P_0^{-1} = \begin{bmatrix} i_\phi & b^T \\ b & I_{xy} \end{bmatrix}. \tag{1.74}$$

Scenario 1: robot stationary

Consider the scenario that the robot is stationary at point **A** and makes n observations.

First assume that the robot can only observe one new feature, feature m. The Jacobian in (1.71) evaluated at the true feature position (x_m, y_m) and the true robot position (x_A, y_A) is denoted as[4]

$$\nabla H_A = [-e \quad -A \quad A], \tag{1.75}$$

where

$$e = \begin{bmatrix} 0 \\ 1 \end{bmatrix}, \quad A = \begin{bmatrix} \frac{dx_A}{r_A} & \frac{dy_A}{r_A} \\ -\frac{dy_A}{r_A^2} & \frac{dx_A}{r_A^2} \end{bmatrix}, \tag{1.76}$$

with

$$\begin{aligned} dx_A &= x_m - x_A \\ dy_A &= y_m - y_A \\ r_A &= \sqrt{dx_A^2 + dy_A^2}. \end{aligned} \tag{1.77}$$

---

[4] For the theoretical convergence results, the Jacobians are always evaluated at the true states. In the real SLAM applications, the Jacobians have to be evaluated at the estimated states and this may cause inconsistency. A detailed analysis of this is given in Section 1.3.3.

For convenience, further denote that

$$A_e = \begin{bmatrix} A^{-1}e & E \end{bmatrix} \tag{1.78}$$

where E denotes 2×2 identity matrix (see footnote 3).

Theorem 1.5. *If the robot is stationary and observes a new feature n times, the covariance matrix of the robot pose and the new feature position estimates is*

$$P^n_{A_{end}} = \begin{bmatrix} P_0 & P_0 A_e^T \\ A_e P_0 & A_e P_0 A_e^T + \frac{A^{-1}R_A A^{-T}}{n} \end{bmatrix} \tag{1.79}$$

where P_0 is the initial robot uncertainty given in (1.73), A is defined in (1.76), A_e is defined in (1.78), and R_A is the observation noise covariance matrix. In the limit when $n \to \infty$, the covariance matrix becomes

$$P^\infty_{A_{end}} = \begin{bmatrix} P_0 & P_0 A_e^T \\ A_e P_0 & A_e P_0 A_e^T \end{bmatrix} = \begin{bmatrix} E \\ A_e \end{bmatrix} P_0 \begin{bmatrix} E & A_e^T \end{bmatrix}. \tag{1.80}$$

Proof: See Appendix A.2.

The following corollary can be obtained from Theorem 1.5.

Corollary 1.1. *If the robot is stationary and observes a new feature n times, the robot uncertainty remains unchanged. The limit (lower bound) on the covariance matrix of the new feature is*

$$P^\infty_{A_m} = A_e P_0 A_e^T, \tag{1.81}$$

which is determined by the robot uncertainty P_0 and the Jacobian ∇H_A. In the special case when the initial uncertainty of the robot orientation p_ϕ is 0, $P^\infty_{A_m}$ is equal to the initial robot position uncertainty P_{xy} in (1.73).

Proof: It is clear that the uncertainty of the robot does not change in (1.79) (will always be P_0). The limit $P^\infty_{A_m}$ in (1.81) can be computed further as

$$\begin{aligned} P^\infty_{A_m} &= A_e P_0 A_e^T \\ &= \begin{bmatrix} A^{-1}e & E \end{bmatrix} \begin{bmatrix} p_\phi & p_{xy\phi}^T \\ p_{xy\phi} & P_{xy} \end{bmatrix} \begin{bmatrix} e^T A^{-T} \\ E \end{bmatrix} \\ &= P_{xy} + A^{-1}e p_\phi e^T A^{-T} + A^{-1}e p_{xy\phi}^T + p_{xy\phi} e^T A^{-T}. \end{aligned}$$

When $p_\phi \to 0$ (then $p_{xy\phi} \to 0$ because P_0 is positive definite), the limit $P^\infty_{A_m} \to P_{xy}$. The proof is completed.

Remark 1.2. Theorem 1.5 and Corollary 1.1 can be regarded as the nonlinear version of Theorem 1.3 (Theorem 3 in [Dissanayake *et al.* (2001)]). Moreover, it is clear that the robot orientation uncertainty has a significant effect on the limit of the feature uncertainty. "*When the robot position is exactly known but its orientation is uncertain, even if there is a perfect knowledge about the relative location between the feature and the robot, it is still impossible to tell exactly where the true feature position is*".

Figures 1.6(a) and 1.6(b) show that the initial robot orientation uncertainty has a significant effect on the feature estimation accuracy. In Figure 1.6(a), the initial uncertainty of the robot pose is $P_0 = diag(0.03, 1, 1)$. Because the robot orientation uncertainty is large (the standard deviation is 0.1732 radians $\approx$ 10 degrees), in the limit, the uncertainty of the feature position is much larger than the initial uncertainty of the robot position. In Figure 1.6(b), the initial robot pose uncertainty is $P_0 = diag(0.001, 1, 1)$. Because the robot orientation uncertainty is very small (the standard deviation is 0.0316 radians $\approx$ 1.8 degrees), in the limit, the uncertainty of the feature position is very close to the initial uncertainty of the robot position.

Now suppose the robot can observe two new features (feature m and feature $\bar{m}$) at point **A**, then the dimension of the observation function in (1.64) is four (two ranges and two bearings), the Jacobian can be denoted as:

$$\nabla \hat{H}_A = \begin{bmatrix} -e & -A & A & 0 \\ -e & -\bar{A} & 0 & \bar{A} \end{bmatrix} \tag{1.82}$$

where $\bar{A}$ is similar to A in (1.76) but defined for feature $\bar{m}$.

Similar to (1.78), denote

$$\bar{A}_e = \begin{bmatrix} \bar{A}^{-1} e & E \end{bmatrix}. \tag{1.83}$$

The following theorem and corollary can now be obtained. The proofs are similar to that of Theorem 1.5 and Corollary 1.1 and are omitted here.

Theorem 1.6. *If the robot is stationary and observes two new features n times, the covariance matrix of the robot pose and the two new feature position estimates is*

$$\hat{P}^n_{A_{end}} = \begin{bmatrix} P_0 & P_0 A_e^T & P_0 \bar{A}_e^T \\ A_e P_0 & P^n_{A_m} & A_e P_0 \bar{A}_e^T \\ \bar{A}_e P_0 & \bar{A}_e P_0 A_e^T & P^n_{\bar{A}_{\bar{m}}} \end{bmatrix} \tag{1.84}$$

where

$$
\begin{aligned}
P^n_{A_m} &= A_e P_0 A_e^T + \frac{A^{-1} R_A A^{-T}}{n}, \\
P^n_{\bar{A}_{\bar{m}}} &= \bar{A}_e P_0 \bar{A}_e^T + \frac{\bar{A}^{-1} R_{\bar{A}} \bar{A}^{-T}}{n},
\end{aligned} \tag{1.85}
$$

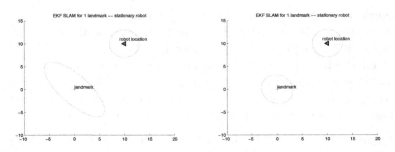

(a) Initial robot orientation uncertainty is large

(b) Initial robot orientation uncertainty is small

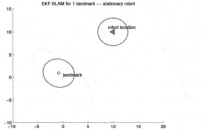
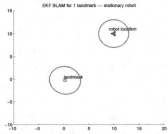

(c) Inconsistency of EKF SLAM

(d) Inconsistency can be neglected when initial robot orientation uncertainty is small

Fig. 1.6   The limits of feature uncertainty when the robot is stationary and observes the feature $n \to \infty$ times (see Theorem 1.5, Corollary 1.1 and Theorem 1.9): In Figure 1.6(a) and Figure 1.6(c), the initial uncertainty of the robot pose is $P_0 = diag(0.03, 1, 1)$. In Figure 1.6(b) and Figure 1.6(d), the initial robot pose uncertainty is $P_0 = diag(0.001, 1, 1)$. For Figure 1.6(a) and Figure 1.6(b), the Jacobians are evaluated at the true robot and feature locations. In Figures 1.6(c) and 1.6(d), the solid ellipses are the limit of the uncertainties when the Jacobians are evaluated at the updated state estimate at each update step.

and $R_{\bar{A}}$ is the observation noise covariance matrix for observing feature $\bar{m}$. In the limit when $n \to \infty$, the whole covariance matrix is

$$\hat{P}^{\infty}_{\bar{A}_{end}} = \begin{bmatrix} P_0 & P_0 A_e^T & P_0 \bar{A}_e^T \\ A_e P_0 & A_e P_0 A_e^T & A_e P_0 \bar{A}_e^T \\ \bar{A}_e P_0 & \bar{A}_e P_0 A_e^T & \bar{A}_e P_0 \bar{A}_e^T \end{bmatrix}$$

$$= \begin{bmatrix} E \\ A_e \\ \bar{A}_e \end{bmatrix} P_0 \begin{bmatrix} E & A_e^T & \bar{A}_e^T \end{bmatrix}. \tag{1.86}$$

Corollary 1.2. *If the robot is stationary and observes two new features* n *times, the robot uncertainty remains unchanged. The limit (lower bound) of the covariance matrix associated with the two new features is*

$$P^{\infty}_{A_{m\bar{m}}} = \begin{bmatrix} A_e P_0 A_e^T & A_e P_0 \bar{A}_e^T \\ \bar{A}_e P_0 A_e^T & \bar{A}_e P_0 \bar{A}_e^T \end{bmatrix}. \tag{1.87}$$

In the special case when the initial uncertainty of the robot orientation $p_\phi = 0$, *the limit* $P^{\infty}_{A_{m\bar{m}}} = \begin{bmatrix} P_{xy} & P_{xy} \\ P_{xy} & P_{xy} \end{bmatrix}.$

Remark 1.3. Theorem 1.6 and Corollary 1.2 are the analogue of Theorem 2 in [Dissanayake *et al.* (2001)]. However, because $A_e \neq \bar{A}_e$, $A_e P_0 A_e^T \neq \bar{A}_e P_0 \bar{A}_e^T$ when $p_\phi \neq 0$. This means that the limits of the uncertainties of the two features are different when the robot orientation uncertainty is not zero. This is different from the linear results proved in [Dissanayake *et al.* (2001)], where the uncertainties of all the features (with similar feature types) are the same. This result is due to the nonlinearity of the observation function, which makes the Jacobians to be different when evaluated at locations of different features.

Figure 1.7(a) shows that the difference between the uncertainties of the two features is large when the robot orientation uncertainty p_ϕ is large (p_ϕ is the same as that in Figure 1.6(a)). Figure 1.7(b) shows that the difference is very small when the initial robot orientation uncertainty p_ϕ is small (p_ϕ is the same as that in Figure 1.6(b)).

Scenario 2: robot moves

Consider the scenario that the robot first remains stationary at point **A** and makes observations $n \to \infty$ times. Then the robot moves to another observation point **B** in one time step, and observes the same features l times.

First assume that the robot can only observe one new feature (at points **A** and **B**) – feature m. The Jacobian in (1.71) evaluated at point **B** and the true position of feature m is denoted as

$$\nabla H_B = [-e \quad -B \quad B], \tag{1.88}$$

where B is similar to A in (1.76) but defined for the robot pose at point **B**. Similar to (1.78), denote

$$B_e = [B^{-1}e \quad E]. \tag{1.89}$$

The following lemma gives the relationship between the Jacobians at point **A** and point **B**.

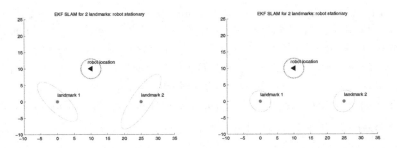

(a) Initial robot orientation uncertainty is large   (b) Initial robot orientation uncertainty is small

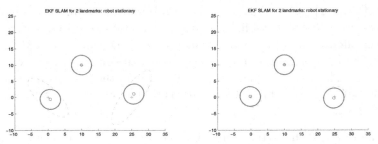

(c) Inconsistency of EKF SLAM for two features   (d) Inconsistency can be neglected when initial robot orientation uncertainty is small

Fig. 1.7   The limits of the two feature uncertainties when the robot is stationary and makes observation $n \to \infty$ times: Figure 1.7(a) shows that the final uncertainties of the two features are different. See Theorem 1.6, Corollary 1.2, Remark 1.3, Theorem 1.9 and the caption of Figure 1.6 for more explanations.

Lemma 1.1. *The relationship between the Jacobians at point* **A** *and point* **B** *is*

$$A_e = B_e \nabla f^A_{\phi X_r}, \qquad (1.90)$$

where $\nabla f^A_{\phi X_r}$ *is the Jacobian of* f *in (1.53) with respect to the robot orientation and position (see equation (1.61)), evaluated at the robot pose* **A** *and the associated control values.*

Proof: See Appendix A.2.

The relationship given in Lemma 1.1 plays an important role in deriving the following convergence results.

Theorem 1.7. *If the robot first remains stationary at point* **A** *and observes one new feature $n \to \infty$ times before it moves to point* **B** *and observes the same feature l times, then the final covariance matrix is*

$$P^l_{B_{end}} = P^0_{B_{start}} + P^l_B \tag{1.91}$$

where

$$P^0_{B_{start}} = \begin{bmatrix} \nabla f^A_{\phi X_r} P_0 (\nabla f^A_{\phi X_r})^T & \nabla f^A_{\phi X_r} P_0 A^T_e \\ A_e P_0 (\nabla f^A_{\phi X_r})^T & A_e P_0 A^T_e \end{bmatrix}$$
$$= \begin{bmatrix} \nabla f^A_{\phi X_r} & 0 \\ 0 & E \end{bmatrix} P^\infty_{A_{end}} \begin{bmatrix} (\nabla f^A_{\phi X_r})^T & 0 \\ 0 & E \end{bmatrix}, \tag{1.92}$$

$$P^l_B = \begin{bmatrix} \nabla f^A_{\gamma v} \Sigma^l_B (\nabla f^A_{\gamma v})^T & 0 \\ 0 & 0 \end{bmatrix}, \tag{1.93}$$

with

$$\Sigma^l_B = [\Sigma^{-1} + l H^T_{AB} R^{-1}_B H_{AB}]^{-1} \geq 0 \tag{1.94}$$

and

$$H_{AB} = [e \quad B] \nabla f^A_{\gamma v}. \tag{1.95}$$

Furthermore, if the matrix $H^T_{AB} R^{-1}_B H_{AB}$ is invertible[5], then the matrix $P^l_B \to 0$ when $l \to \infty$. Here $P^\infty_{A_{end}}$ is defined in (1.80), R_B is the covariance matrix of the observation noise at point **B**, *$\nabla f^A_{\phi X_r}$ and $\nabla f^A_{\gamma v}$ are Jacobians of function f in (1.53) evaluated at point* **A** *and the associated control values.*

Proof: See Appendix A.2.

By Theorem 1.7, the lower bound of the covariance matrix is $P^0_{B_{start}}$, which is the covariance matrix when the robot first reaches point **B** if there is no control noise in moving from **A** to **B** ($\Sigma = 0$ in (1.59)).

Figures 1.8(a), 1.8(b), 1.8(c), and 1.8(d) illustrate Theorem 1.7. The initial robot uncertainty is the same as that used for Figure 1.6(a). Figures 1.8(a) and 1.8(b) show the case when there is no control noise. Figure 1.8(a) shows the uncertainties after the prediction step and Figure 1.8(b) shows

---

[5]This depends on the process model and the direction of the robot movement but this is true in most of the cases.

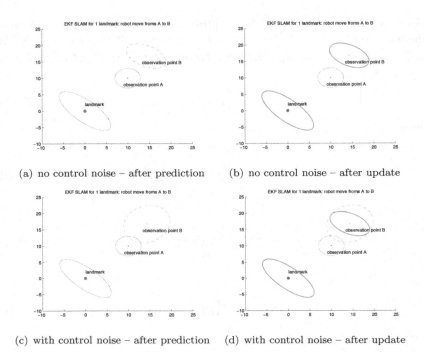

(a) no control noise – after prediction   (b) no control noise – after update

(c) with control noise – after prediction   (d) with control noise – after update

Fig. 1.8   The limits of the robot and feature uncertainties when the robot first remains stationary at point **A** and makes observation $n \to \infty$ times and then moves to **B** and observes the same feature $l \to \infty$ times (see Theorem 1.7).

the uncertainties after the update using the observations at point **B**. It can be seen that the observations at point **B** can not reduce the uncertainty of the robot and features. Figures 1.8(c) and 1.8(d) show the case when control noise is present. In this case, the feature uncertainty cannot be improved by the observation at point **B**, while the uncertainty of the robot can be reduced to the same level as the case when there is no control noise. The limits of the uncertainties are independent of the extent of sensor and control noises. The control noise only affect the robot uncertainty after the prediction in Figure 1.8(c). The sensor noise used are the same as those in Figure 1.6, the robot speed and the control noises (in Figures 1.8(c) and 1.8(d)) are deliberately enlarged, just to make the differences of the ellipses visible.

Now suppose the robot can observe two new features (feature m and feature $\bar{m}$) at points **A** and **B**, then the dimension of the observation function in (1.64) is four (two ranges and two bearings), denote the corresponding

Jacobians as $\nabla \hat{H}_A$ given in (1.82) and

$$\nabla \hat{H}_B = \begin{bmatrix} -e & -B & B & 0 \\ -e & -\bar{B} & 0 & \bar{B} \end{bmatrix}. \tag{1.96}$$

Theorem 1.8. *If the robot first remains stationary at point* **A** *and observes two new features* $n \to \infty$ *times before it moves to point* **B** *and observes the same two features* l *times, then the final covariance matrix is*

$$\hat{P}^l_{B_{end}} = \hat{P}^0_{B_{start}} + \hat{P}^l_B \tag{1.97}$$

where

$$\hat{P}^0_{B_{start}} = \begin{bmatrix} \nabla f^A_{\phi X_r} & 0 & 0 \\ 0 & E & 0 \\ 0 & 0 & E \end{bmatrix} \hat{P}^\infty_{A_{end}} \begin{bmatrix} (\nabla f^A_{\phi X_r})^T & 0 & 0 \\ 0 & E & 0 \\ 0 & 0 & E \end{bmatrix} \tag{1.98}$$

$$\hat{P}^l_B = \begin{bmatrix} \nabla f^A_{\gamma v} \hat{\Sigma}^l_B (\nabla f^A_{\gamma v})^T & 0 & 0 \\ 0 & 0 & 0 \\ 0 & 0 & 0 \end{bmatrix}, \tag{1.99}$$

with

$$\hat{\Sigma}^l_B = [\Sigma^{-1} + l(H^T_{AB} R^{-1}_B H_{AB} + H^T_{A\bar{B}} R^{-1}_{\bar{B}} H_{A\bar{B}})]^{-1} \geq 0, \tag{1.100}$$

and

$$H_{A\bar{B}} = [e \;\; \bar{B}] \nabla f^A_{\gamma v}. \tag{1.101}$$

Furthermore, if the matrix $H^T_{AB} R^{-1}_B H_{AB} + H^T_{A\bar{B}} R^{-1}_{\bar{B}} H_{A\bar{B}}$ *is invertible, then the matrix* $\hat{P}^l_B \to 0$ *when* $l \to \infty$. *Here* $\hat{P}^\infty_{A_{end}}$ *is defined in (1.86),* $\nabla f^A_{\phi X_r}$ *and* $\nabla f^A_{\gamma v}$ *are Jacobians of function* f *in (1.53) evaluated at point* **A** *and the associated control values,* H_{AB} *is defined in (1.95), and* $R_B, R_{\bar{B}}$ *are the covariance matrices of the observation noise at point* **B** *and for observing features* m *and* $\bar{m}$, *respectively.*

Proof: The proof is similar to that of Theorem 1.7 and is omitted.

Remark 1.4. Theorems 1.7 and 1.8 show that the only effect of the observations made at point **B** is to reduce the additional robot uncertainty generated from the process noise. *The observations made at point* **B** *cannot reduce the uncertainty of the feature further if the robot had already* *observed the feature many times at point* **A**. Theorems 1.7 and 1.8 can be extended to the case when it takes more than one step to move from **A** to **B** such as **A** $\rightarrow$ **B**$_1$ $\rightarrow$ **B**$_2$ $\rightarrow$ ... $\rightarrow$ **B**$_n$ $\rightarrow$ **B**. For example, for the one feature case, the limit of the covariance matrix satisfies

$$P^{\infty}_{B_{end}} \geq \begin{bmatrix} F_{AB}P_0F^T_{AB} & F_{AB}P_0A^T_e \\ A_eP_0F^T_{AB} & A_eP_0A^T_e \end{bmatrix}, \tag{1.102}$$

where

$$F_{AB} = \nabla f^{B_n}_{\phi X_r} \cdots \nabla f^{B_1}_{\phi X_r} \nabla f^{A}_{\phi X_r}. \tag{1.103}$$

Figures 1.9(b) and 1.9(d) illustrate the results.

In summary, most of the convergence properties proved in [Dissanayake *et al.* (2001)] can be generalized to practical nonlinear SLAM problems, provided that all the Jacobians are evaluated at the true values. Under this condition, the explicit formulas for the covariance matrices can also be achieved for several basic scenarios. In practice, since the Jacobians are evaluated at the estimated state values, the covariance matrices obtained may be slightly different from the ideal ones. When the linearization error is large, EKF SLAM algorithm may also provide significantly inconsistent estimate, as shown in the next section.

1.3.3   *EKF SLAM Consistency*

This section discusses the consistency of EKF based SLAM algorithms. An estimator is called consistent if the actual mean square error matches the calculated covariance.

In the past few years, a number of researchers have demonstrated that the lower bound for the map accuracy presented in [Dissanayake *et al.* (2001)] is violated and the EKF SLAM produces inconsistent estimates due to errors introduced during the linearization process [Julier and Uhlmann (2001b); Castellanos *et al.* (2004); Frese (2006a); Martinell *et al.* (2005); Bailey *et al.* (2006)]. In [Frese (2006a)] and [Bailey *et al.* (2006)] it was pointed out that the robot orientation uncertainty is the main cause of the inconsistency in EKF SLAM. Although extensive simulation results are available to show that the inconsistency does exist, and almost all of the related papers point out that linearization is the cause of the inconsistency, the only theoretical explanation is given by [Julier and Uhlmann (2001b)]. This work, however, only deals with the case when the robot is stationary.

Further theoretical analysis of the consistency of EKF SLAM can be performed based on the convergence results in the previous section. This can provide some insights into the possible inconsistencies of EKF SLAM that have been recently observed. In fact, for all the theoretical convergence properties proved in the previous section, it is assumed that the Jacobians are evaluated at the true robot pose and the true feature positions. In a real-life SLAM, the true locations of the robot and features are not known, and the Jacobians have to be evaluated at the estimated values. This section proves that this may result in over-confident (inconsistent) estimates. It is also shown that the robot orientation uncertainty plays an important role in the possible inconsistency of EKF SLAM.

1.3.3.1 *Why inconsistency can occur in EKF SLAM?*

A number of recent publications indicate that the key source of EKF SLAM inconsistency is the error introduced during the linearization process (e.g. [Castellanos *et al.* (2004)] and [Frese (2006a)]). While it is clear that linearization is an approximation which can introduce errors into the estimation process, it is reasonable to expect that the incorrect estimate is likely to be either too optimistic (estimated uncertainty smaller than true uncertainty) or too pessimistic (estimated uncertainty larger than true uncertainty). However, the SLAM literature only reports estimator inconsistency as a result of optimistic estimates. No instances of pessimistic estimates during EKF SLAM has been reported. Why?

1.3.3.2 *An intuitive explanation*

In most cases, the measurement available for use in the SLAM algorithms is the relative location between the robot and features, and the objective of the SLAM process is to estimate the absolute robot and feature locations.

Suppose x, y are two numbers and that two noisy measurements of $x - y$ are available:

$$\begin{aligned} x - y &\approx 99.8 \\ x - y &\approx 100.1. \end{aligned} \tag{1.104}$$

Based on these measurements, although it is possible to say "$x - y$ is around 100", nothing can be said about the true values of x and/or y. However, if the measurement equation is non-linear, the linearized version of this equation may look like

$$1.01x - y \approx 99.8$$
$$0.99x - y \approx 100.1. \tag{1.105}$$

From these two equations, the (approximate) absolute values of x and y can be calculated. Obviously, this outcome is too optimistic (inconsistent).

In the next two subsections, it will be shown that the mechanism that causes overconfident estimates in SLAM is similar to that presented above. Theoretical proofs are given for two basic scenarios. Some of the lengthy proofs are given in Appendix A.3.

1.3.3.3   *Scenario 1: robot stationary*

In EKF SLAM, the observation innovation ($\mu(k + 1)$ in (1.68)) is used to update the previous estimate $\hat{X}$. Through linearization, the innovation can be expressed as (see (1.75))

$$\begin{aligned}
\mu &= z - H(\hat{X}) \\
&\approx H(X) - H(\hat{X}) \\
&\approx \nabla H_A (X - \hat{X}) \\
&= -e(\phi - \hat{\phi}) - A(X_r - \hat{X}_r) + A(X_m - \hat{X}_m),
\end{aligned} \tag{1.106}$$

where $\hat{\phi}, \hat{X}_r$ and $\hat{X}_m$ are the estimates of the robot orientation, the robot position, and the feature position, respectively. (1.106) is equivalent to

$$e\phi + AX_r - AX_m \approx -\mu + e\hat{\phi} + A\hat{X}_r - A\hat{X}_m. \tag{1.107}$$

Suppose the robot is stationary at point **A** and makes two consecutive observations to feature m, z_1 and z_2. After the update using z_1, the estimates of the robot orientation, the robot position, and the feature position will change from $\hat{\phi}, \hat{X}_r, \hat{X}_m$ to $\hat{\phi}^1, \hat{X}_r^1, \hat{X}_m^1$, thus the Jacobian will be evaluated at a different point in the state space when z_2 is used for the next update. The two innovations μ_1, μ_2 give

$$\begin{aligned}
e\phi + \tilde{A}_1 X_r - \tilde{A}_1 X_m &\approx -\mu_1 + e\hat{\phi} + \tilde{A}_1 \hat{X}_r - \tilde{A}_1 \hat{X}_m, \\
e\phi + \tilde{A}_2 X_r - \tilde{A}_2 X_m &\approx -\mu_2 + e\hat{\phi}^1 + \tilde{A}_2 \hat{X}_r^1 - \tilde{A}_2 \hat{X}_m^1,
\end{aligned} \tag{1.108}$$

where $\tilde{A}_1, \tilde{A}_2$ are defined in a manner similar to (1.76) but computed at the estimated robot and feature locations. Both $\tilde{A}_1, \tilde{A}_2$ are non-singular matrices that are different but close to A.

The above two equations are equivalent to

$$\begin{aligned}
\tilde{A}_1^{-1} e\phi + X_r - X_m &\approx -\tilde{A}_1^{-1}\mu_1 + \tilde{A}_1^{-1} e\hat{\phi} + \hat{X}_r - \hat{X}_m, \\
\tilde{A}_2^{-1} e\phi + X_r - X_m &\approx -\tilde{A}_2^{-1}\mu_2 + \tilde{A}_2^{-1} e\hat{\phi}^1 + \hat{X}_r^1 - \hat{X}_m^1.
\end{aligned} \tag{1.109}$$

So

$$(\tilde{A}_1^{-1}e - \tilde{A}_2^{-1}e)\phi \approx \tilde{A}_2^{-1}\mu_2 - \tilde{A}_1^{-1}\mu_1 + \hat{X}_r - \hat{X}_m \\ + \tilde{A}_1^{-1}e\hat{\phi} - \tilde{A}_2^{-1}e\hat{\phi}^1 - \hat{X}_r^1 + \hat{X}_m^1. \tag{1.110}$$

By the special structure of $\tilde{A}_1, \tilde{A}_2$ (see (1.76)), if $\tilde{A}_1 \neq \tilde{A}_2$, then $\tilde{A}_1^{-1}e \neq \tilde{A}_2^{-1}e$ and equation (1.110) provides some information on the value of ϕ. It is obvious that *observing a single new feature will not improve the knowledge of the robot orientation.* Therefore, this apparent information on the robot orientation is incorrect and will result in overconfident estimates (inconsistency).

To examine the extent of the possible inconsistency, let the robot be stationary at point **A** and observe a new feature n times. Let the estimate be updated after each observation using Jacobians evaluated at the updated estimate at each time step. Denote the different Jacobians as

$$\nabla H_{\tilde{A}_j} = \begin{bmatrix} -e & -\tilde{A}_j & \tilde{A}_j \end{bmatrix}, \quad 1 \leq j \leq n. \tag{1.111}$$

Let R_A denote the observation noise covariance matrix at point **A**, and define

$$w(n, A) = ne^T R_A^{-1}e - e^T R_A^{-1}(\sum_{j=1}^n \tilde{A}_j) \\ \cdot (\sum_{j=1}^n \tilde{A}_j^T R_A^{-1} \tilde{A}_j)^{-1}(\sum_{j=1}^n \tilde{A}_j^T)R_A^{-1}e. \tag{1.112}$$

As before, suppose that the initial robot uncertainty is P_0 given by (1.73).

Theorem 1.9. *In EKF SLAM, if the robot is stationary at point* **A** *and observes a new feature* n *times, the inconsistency occurs due to the fact that Jacobians are evaluated at different state estimates. The level of inconsistency is determined by the initial robot uncertainty* P_0 *and the* $w(n, A)$ *defined in (1.112). When* $n \to \infty$, *the inconsistency may cause the variance of the robot orientation estimate to be reduced to zero.*

Proof: See Appendix A.3.

Figures 1.6(c), 1.6(d), 1.7(c), and 1.7(d) illustrate the results in Theorem 1.9. In Figure 1.6(c), the initial uncertainty of the robot pose is the same as that used in Figure 1.6(a), the solid ellipse is the limit of the feature uncertainty when the Jacobian is evaluated at the updated state estimate at each update step. This figure is generated by performing 1000 updates assuming that the range and bearing measurements are corrupted by random Gaussian noise (the standard deviations of range and bearing noise are selected to be similar to that of a typical indoor laser scanner, $0.1m$ and 1°, respectively). It can be seen that the uncertainty of the feature is reduced far below the theoretical limit (dashed ellipse), demonstrating the inconsistency of EKF SLAM solution. In Figure 1.6(d), the initial uncertainty of the robot orientation is much smaller (the same as that used in Figure 1.6(b)). It can be seen that the extent of inconsistency is too small to be seen (the solid ellipse almost coincides with the dashed one).

1.3.3.4 *Scenario 2: robot moves*

Consider the scenario that the robot observes a new feature at point **A** and then moves to point **B** and makes an observation of the same feature. Similar to (1.109), the two innovations μ_A, μ_B give

$$\begin{aligned} \tilde{A}^{-1}e\phi^A + X_r^A - X_m &\approx -\tilde{A}^{-1}\mu_A + \tilde{A}^{-1}e\hat{\phi}^A + \hat{X}_r^A - \hat{X}_m^A, \\ \tilde{B}^{-1}e\phi^B + X_r^B - X_m &\approx -\tilde{B}^{-1}\mu_B + \tilde{B}^{-1}e\hat{\phi}^B + \hat{X}_r^B - \hat{X}_m^B. \end{aligned} \quad (1.113)$$

From the process model (1.53) with appropriate linearization,

$$\begin{aligned} \phi^B &\approx \phi^A + f_\phi(\hat{\gamma}, \hat{v}, 0, 0), \\ X_r^B &\approx X_r^A + \begin{bmatrix} \hat{v}T\cos(\hat{\phi}_A) \\ \hat{v}T\sin(\hat{\phi}_A) \end{bmatrix} + \begin{bmatrix} -\hat{v}T\sin(\hat{\phi}_A) \\ \hat{v}T\cos(\hat{\phi}_A) \end{bmatrix}(\phi^A - \hat{\phi}_A). \end{aligned} \quad (1.114)$$

Thus

$$\begin{aligned} &\left(\tilde{A}^{-1}e - \tilde{B}^{-1}e - \begin{bmatrix} -\hat{v}T\sin(\hat{\phi}_A) \\ \hat{v}T\cos(\hat{\phi}_A) \end{bmatrix} \right)\phi^A \\ &\approx \tilde{B}^{-1}\mu_B - \tilde{A}^{-1}\mu_A + \tilde{A}^{-1}e\hat{\phi}^A + \hat{X}_r^A - \hat{X}_m^A - \tilde{B}^{-1}e\hat{\phi}^B - \hat{X}_r^B + \hat{X}_m^B \\ &\quad + \tilde{B}^{-1}ef_\phi(\hat{\gamma}, \hat{v}, 0, 0) + \begin{bmatrix} \hat{v}T\cos(\hat{\phi}_A) \\ \hat{v}T\sin(\hat{\phi}_A) \end{bmatrix} - \begin{bmatrix} -\hat{v}T\sin(\hat{\phi}_A) \\ \hat{v}T\cos(\hat{\phi}_A) \end{bmatrix}\hat{\phi}_A. \end{aligned}$$

$$(1.115)$$

If

$$\tilde{A}^{-1}e \neq \tilde{B}^{-1}e + \begin{bmatrix} -\hat{v}T\sin(\hat{\phi}_A) \\ \hat{v}T\cos(\hat{\phi}_A) \end{bmatrix},$$

then equation (1.115) contains information on ϕ^A, which is clearly incorrect as observations to a single feature do not provide any knowledge about the robot orientation.

Note that

$$\tilde{A}^{-1}e = \tilde{B}^{-1}e + \begin{bmatrix} -\hat{v}T\sin(\hat{\phi}_A) \\ \hat{v}T\cos(\hat{\phi}_A) \end{bmatrix}$$

is actually the relationship proved in Lemma 1.1. Therefore, the following result can now be stated.

Theorem 1.10. *When the robot observes the same feature at two different points* **A** *and* **B**, *the EKF SLAM algorithm may provide inconsistent estimates due to the fact that the Jacobians evaluated at the estimated robot positions may violate the key relationship between the Jacobians as shown in Lemma 1.1.*

Proof: See Appendix A.3.

Figures 1.9(a)-1.9(d) illustrate the extent of inconsistency under scenario 2. The robot first keeps still at point **A** and makes $n = 10000$ observations. The initial robot uncertainty is the same as that used in Figure 1.6(a). The true Jacobians are used at point **A** to guarantee the consistency of the estimate before the robot moves. The robot then moves 500 steps to **B** and keeps observing the same feature while moving. The thin/solid ellipses illustrate the estimate uncertainty after the observation at point **A**. The dashed ellipses correspond to the uncertainties at the intermediate points (every 100 steps) while the thick/solid ellipses illustrate the final uncertainty. Figure 1.9(a) shows that the extent of inconsistency is quite significant when there is no control noise. Figure 1.9(b) shows the corresponding results where true Jacobians are used. Figure 1.9(c) shows the inconsistency when control noise is present. Figure 1.9(d) shows the corresponding results where true Jacobians are used. In this simulation, the sensor noise used were the same as that used in Figure 1.6, the control noise were chosen to be similar to that of Pioneer robots — standard deviations of velocity noise and turn rate noise are $0.02m/s$ and $3^o/s$, respectively. The similarity between Figures 1.9(b) and 1.9(d) is due to the relatively small sensor noise where after the update, the uncertainty is almost the same as that obtained when there is no control noise (see Figure 1.8).

In the simulations presented here, the magnitudes of the sensor noise and control noise were selected to be similar to those of a typical indoor-laser and Pioneer robots (except for the control noise in Figure 1.8). The effects of the sensor noise and control noise on the extent of inconsistency are complex and need further investigation. In general, larger noise may result in larger errors in the Jacobians but the amount of "wrong information" contained in (1.110) or (1.115) is also less when the noise is larger.

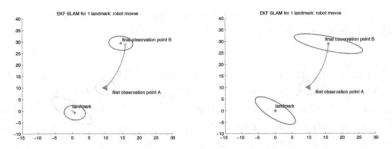

(a) move 500 steps with no control noise – inconsistency

(b) move 500 steps with no control noise – using true Jacobians

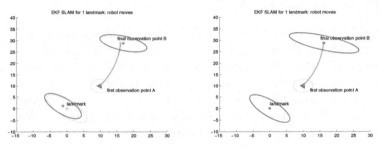

(c) move 500 steps with control noise – inconsistency

(d) move 500 steps with control noise – using true Jacobians

Fig. 1.9   The inconsistency of EKF SLAM when the robot moves (see Theorem 1.10): The robot first remains stationary at point **A** and makes observation $n = 10000$ times. Then the robot moves 500 steps to **B** and keeps observing the same feature while moving. The thin/solid ellipses are the uncertainty after the observation at point **A**, the dashed ellipses are the uncertainties at the intermediate points (every 100 steps), the thick/solid ellipses are the final uncertainties.

The inconsistency results presented here only focus on the covariance matrices. The inconsistent mean estimate naturally results from the inconsistent covariance matrix because the Kalman gain in the subsequent step will be incorrect once the covariance matrix becomes inconsistent. See for example the means in Figures 1.6(c), 1.7(c), 1.9(a), 1.9(c).

In the results presented in this section, the consistency of SLAM estimate is evaluated based on the fact "keep observing new features does not help in reducing the robot pose uncertainty". The inconsistency is evidenced by "incorrect reduction of the covariance matrix of the robot pose estimate". Similar approaches have been used in [Julier and Uhlmann (2001b); Castellanos *et al.* (2004); Bailey *et al.* (2006)].

In summary, the results in this section show that in EKF SLAM, inconsistent estimates can be produced when the Jacobians are evaluated using the estimated states, as the case in practice. It is also shown that when the robot orientation uncertainty is large, the extent of inconsistency is significant; when the robot orientation uncertainty is small, the extent of inconsistency is insignificant.

The insights on the fundamental reasons why EKF SLAM can be inconsistent can help in deriving new variations of SLAM algorithms that minimize the extent of possible inconsistency. For example, since the robot orientation error is one of the main causes of EKF SLAM inconsistency, the use of local submaps, in which the robot orientation uncertainties are kept very small, can improve consistency. Moreover, if smoothing is applied such as in SAM [Dellaert and Kaess (2006)], Iterated Sparse Local Submap Joining Filter [Huang *et al.* (2008b)] and Iterated D-SLAM Map Joining [Huang *et al.* (2009)], the consistency can be further improved.

1.4 Motivation

EKF based SLAM has been investigated for many years, and the principle has been well understood. In recent years, properties of EIF based SLAM approaches have also been intensively studied. From the practical point of view, representations of the environment are expected in an efficient manner or real-time. As such, EIF based approaches are promising as they capture the sparse structure of the problem.

It is known that the information matrix associated with SLAM contains a limited number of dominating elements, whereas all other elements are near zero when the information matrix is properly normalized, indicating a promising direction to pursue in the quest for obtaining computationally efficient SLAM solutions. The sparse representation for solving SLAM problems has drawn significant attention since the development of the Sparse Extended Information Filter (SEIF) by [Thrun *et al.* (2004a)]. It has been shown that significant computational advantages can be achieved by exploiting the sparseness of the information matrix or by using techniques from sparse graph and sparse linear algebra (e.g. [Dellaert and Kaess (2006); Eustice *et al.* (2006); Thrun *et al.* (2005)]). Frese (2005) proved that the off-diagonal elements in the information matrix of the conventional EIF SLAM decay exponentially with the distance the robot traveled between seeing the pair of related features. This provides the theoretical foundation for sparse information filters and other methods which exploit the sparse representation such as the Thin Junction Tree filter [Paskin (2003)] and the Treemap algorithm [Frese (2006b)].

While the direction of exploiting the sparse representation to solve the SLAM problem is promising, there exist a number of practical issues that still need addressing. It is known that the near-zero elements in the information matrix can not be simply set to zero as this will cause inconsistent estimates [Eustice *et al.* (2005b)]. Existing techniques for generating the sparse information matrix result in either an increase in the size of the state vector or information loss. Data association and recovery of the state estimate and the associated covariances require significant computational effort and need efficient solutions for EIF-based SLAM to be practical.

This book presents a further investigation of the sparse information filter based SLAM. Mechanisms that cause the information matrix to be exactly sparse are studied and a decoupled SLAM algorithm (D-SLAM) is presented, in which the localization and mapping are decoupled into two concurrent yet separate processes. With the robot pose removed from the state vector for mapping, this algorithm avoids the operation of marginalizing out the past robot pose from the state vector in each prediction step which makes the information matrix dense [Eustice *et al.* (2006); Frese (2005)]. Thus an exactly sparse information matrix for mapping is achieved. Two efficient recovery methods are developed based on the observation that the information matrix evolves gradually during the estimation process.

Even with computational advantages present in sparse information filter based solutions, the size of the environment for which SLAM is practical is still limited. Local map based solutions have addressed SLAM in large scale environments. In Local Map Sequencing [Tardos *et al.* (2002)] and Constrained Local Submap Filter (CLSF) [Williams (2001)], local maps are built with a small scale traditional EKF SLAM algorithm. These local maps are then combined into a large scale global map. In the Atlas system [Bosse *et al.* (2004)] and the network coupled feature maps (NCFM) algorithm [Bailey (2002)], the local maps are organized in a global topological graph. The nodes of this graph represent local maps whereas the edges represent the transformation or relative location between adjacent local maps. In Hierarchical SLAM [Estrada *et al.* (2005)], the relative location of two consecutive local maps is maintained by estimating the transformation of the two robot poses which represent the origins of the coordinate frame attached to these local maps. On the global level, a relative stochastic map containing transformations among local map coordinate frames is used to estimate the relative location of all the local maps. Many of the existing local map based SLAM algorithms use an EKF framework and the global map update is costly in computation.

This book presents two methods combining the local map formulation and the sparse information filter framework for large scale SLAM. The first method, the D-SLAM Local Map Joining Filter, extends the D-SLAM strategy to allow joining local maps; and the second method, Sparse Local Submap Joining Filter (SLSJF), incorporates a limited set of robot poses into the global state vector. In both methods, the local maps are treated as virtual observations to a set of features. The state vector of the global map is structured such that the marginalization operations are avoided and that the information matrix of the global map becomes exactly sparse. It will be shown that significant savings in computational effort can be obtained. In SLSJF, the sparse information matrix is achieved without incurring any information loss apart from the linearization process, which exists in all EKF/EIF based approaches.

1.5 Book Overview

The main contributions of this book are as follows:

- It is shown that the decoupling of localization and mapping is possible while using an absolute map without any redundant elements present in the state vector. This is achieved by recasting the range and bearing observations from the robot to features in the environment into a new equivalent observation, which contains information about the relative locations among the features.
- It is demonstrated that the new decoupled SLAM formulation naturally results in an exactly sparse information matrix for mapping. Eliminating near-zero elements is not required to achieve sparseness. Furthermore, it is shown that the information matrix retains its sparseness even after loop closures and that the extent of sparseness is related to the range of the sensor on board the robot and the feature density in the environment.
- Two state and covariance recovery methods are developed. One is based on the Preconditioned Conjugated Gradient algorithm that incorporates an incremental preconditioning process. The other uses a direct linear equation solving technique based on an incre-

mental complete Cholesky Factorization process. These methods exploit the similarity of the information matrix in two consecutive time steps to achieve low computational cost.

- The D-SLAM Local Map Joining Filter is developed for the mapping of large scale environments by combining local maps using the D-SLAM framework. Relative location information among the features in the local map is extracted and fused by an EIF to build a global map. The resulting information matrix of the global map is exactly sparse.

- The Sparse Local Submap Joining Filter (SLSJF) is developed by directly fusing the local map estimate into the global map which contains all feature locations and the robot start/end poses in each local map. The resulting information matrix of the global map is exactly sparse. The information loss caused by D-SLAM and the D-SLAM Local Map Joining Filter is completely avoided.

The book is organized as follows:

Chapter 2   analyzes the evolution of the information matrix under the filter operations in EIF SLAM. The implication of conditional independence among the states is studied. Two conditions that are necessary and sufficient to achieve the exact sparseness of the information matrix are described and the strategies that have been exploited by the algorithms presented in this book are examined in the context of these two conditions.

Chapter 3   describes the D-SLAM algorithm in which the SLAM process is reformulated as two concurrent yet separate processes for localization and mapping. The processes of mapping and localization are described in detail. It is also shown that the information matrix associated with mapping in D-SLAM is exactly sparse. Two efficient methods for recovering the state estimate and the associated covariances are described.

Chapter 4   presents the D-SLAM Local Map Joining Filter for the mapping of large scale environments. The processes of building local maps, extracting relative location information and fusing the local maps into the global map using the D-SLAM framework are described in detail.

Chapter 5   describes the Sparse Local Submap Joining Filter (SLSJF). It is demonstrated that by maintaining a global state vector including all feature locations as well as the robot start and end poses in each local map, the SLSJF completely avoids the information loss which is present in the algorithms described in Chapters 3 and 4.

Appendix A presents proofs for some results of EKF SLAM convergence and consistency properties.

Appendix B provides the incremental method for Cholesky Factorization of the information matrix in SLAM.

Chapter 2

Sparse Information Filters in SLAM

Recent research has shown that the information matrix of the SLAM problem can be made sparse, i.e. a substantial amount of its off-diagonal elements can be made equal to zero. The sparseness of the information matrix is in contrast to the dense structure of its inverse, the covariance matrix. It has been proved by [Frese (2005)] that the off-diagonal elements of the information matrix in the conventional EIF SLAM decay exponentially with the distance the robot travelled between seeing the pair of features that correspond to the respective elements. This provides the theoretical foundation for the sparse representation in SLAM. The update in EIF is additive [Manyika and Durrant-Whyte (1994)] and is a constant time operation [Thrun *et al.* (2004a)], provided that the state estimate is available for computing Jacobians and the number of features present in the observation is bounded. The sparse structure of the information matrix can also make the prediction step efficient [Thrun *et al.* (2004a)]. The significant computational savings that can be achieved has motivated a number of SLAM algorithms using the EIF.

This chapter analyzes the evolution of the information matrix in EIF SLAM and the meaning of the zero off-diagonal elements in it. The conditions that need to be satisfied in order to achieve the exact sparseness are examined and the strategies that have been exploited by the algorithms in this book are also explained.

2.1 Information Matrix in the Full SLAM Formulation

In this section, the evolution of the information matrix in the full SLAM formulation using the EIF will be analyzed in detail. This formulation is defined as the one in which all robot poses and observed features are present in the state vector. Three operations on the information matrix are required during the SLAM process including: initializing a new robot pose, initializing a new feature and updating using observations.

A simple $2D$ range-bearing SLAM example as shown in Figure 2.1 is used for the analysis. In this example, the robot starts from the origin and moves to P_1 where two features, f_1 and f_2, are observed for the first time. Then the robot moves to P_2 where f_2 and a new feature, f_3, are observed. Normally in SLAM implementations, the robot makes some observations before it starts moving. However, in this example, to facilitate the analysis, the robot first moves away from the origin and then makes observations.

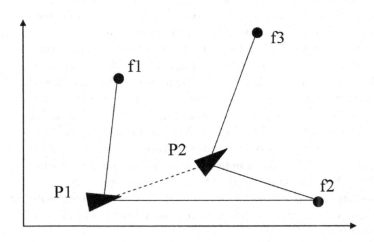

Fig. 2.1   2D SLAM example.

Let the states for the robot poses at P_1 and P_2 be X_{r1} and X_{r2} respectively, and the states for features f_1, f_2 and f_3 be X_{m1}, X_{m2} and X_{m3} respectively. Suppose the initial information matrix is a 3×3 matrix for X_{r1}, which incorporates the process information from the coordinate origin to P_1.

Denote the two range-bearing observations made at P_1 about f_1 and f_2 by

$$Z_1 = \begin{bmatrix} z_1 \\ z_2 \end{bmatrix} = \begin{bmatrix} h(X_{r1}, X_{m1}) + v_1 \\ h(X_{r1}, X_{m2}) + v_2 \end{bmatrix} \qquad (2.1)$$

where h represents the range-bearing observation model in (1.6), and the observation noises are denoted as v_1 and v_2 assumed to be white Gaussian with zero mean. The covariance matrix of Z_1 is denoted as R_1.

The information matrix is first expanded by adding zeros for features f_1 and f_2, and the state vector and the information matrix are

$$\begin{bmatrix} X_{r1} \\ X_{m1} \\ X_{m2} \end{bmatrix}, \quad \begin{bmatrix} *_{(3\times3)} & 0_{(3\times2)} & 0_{(3\times2)} \\ 0_{(2\times3)} & 0_{(2\times2)} & 0_{(2\times2)} \\ 0_{(2\times3)} & 0_{(2\times2)} & 0_{(2\times2)} \end{bmatrix} \qquad (2.2)$$

where the $*_{(.)}$ represents a non-zero block with its dimension and each $0_{(.)}$ represents a zero block with its dimension.

Then the EIF update equation

$$I_{new} = I_{old} + \nabla H_X^T R^{-1} \nabla H_X \qquad (2.3)$$

is used to update the information matrix. Here I_{old} and I_{new} represent the information matrix before and after the observation is fused respectively; and ∇H_X is the Jacobian of the stacked observation functions with respect to the states.

The resulting state vector and the information matrix are

$$\begin{bmatrix} X_{r1} \\ X_{m1} \\ X_{m2} \end{bmatrix}, \quad \begin{bmatrix} *_{(3\times3)} & *_{(3\times2)} & *_{(3\times2)} \\ *_{(2\times3)} & *_{(2\times2)} & 0_{(2\times2)} \\ *_{(2\times3)} & 0_{(2\times2)} & *_{(2\times2)} \end{bmatrix} \qquad (2.4)$$

where each $*_{(.)}$ represents a non-zero block with its dimension. Note that only the pairs (X_{r1}, X_{m1}) and (X_{r1}, X_{m2}) that appear in the same observation equation have non-zero off-diagonal blocks in the information matrix.

Let the robot motion from P_1 to P_2 be represented by

$$X_{r2} = f_r(X_{r1}, u_1) + w_1 \qquad (2.5)$$

where u_1 is the control measurement and w_1 is the process noise assumed to be white Gaussian with zero mean and a covariance Q_1.

Equation (2.5) can be transformed as

$$
\begin{aligned}
0 &= f_r(X_{r1}, u_1) - X_{r2} + w_1 \\
  &= F(X_{r1}, X_{r2}, u_1) + w_1
\end{aligned}
\tag{2.6}
$$

such that the control measurement can be fused in the same way as the observation using

$$
I'_{new} = I'_{old} + \nabla F_{12}^T Q_1^{-1} \nabla F_{12}
\tag{2.7}
$$

where I'_{old} and I'_{new} represent the information matrix before and after the control measurement is fused respectively, and ∇F_{12} is the Jacobian of the function F with respect to the states.

The resulting state vector and the information matrix are

$$
\begin{bmatrix} X_{r1} \\ X_{r2} \\ X_{m1} \\ X_{m2} \end{bmatrix}, \quad
\begin{bmatrix}
*_{(3\times3)} & *_{(3\times3)} & *_{(3\times2)} & *_{(3\times2)} \\
*_{(3\times3)} & *_{(3\times3)} & 0_{(3\times2)} & 0_{(3\times2)} \\
*_{(2\times3)} & 0_{(2\times3)} & *_{(2\times2)} & 0_{(2\times2)} \\
*_{(2\times3)} & 0_{(2\times3)} & 0_{(2\times2)} & *_{(2\times2)}
\end{bmatrix}
\tag{2.8}
$$

where each $*_{(.)}$ represents a non-zero block with its dimension. Note that X_{r2} only has non-zero off-diagonal blocks with X_{r1} due to their concurrent appearance in the process equation (2.6).

Let the two range-bearing observations made at P_2 be denoted by

$$
Z_2 = \begin{bmatrix} z_3 \\ z_4 \end{bmatrix} = \begin{bmatrix} h(X_{r2}, X_{m2}) + v_3 \\ h(X_{r2}, X_{m3}) + v_4 \end{bmatrix}
\tag{2.9}
$$

where v_3 and v_4 are the observation noise assumed to be white Gaussian with zero mean. The covariance matrix of Z_2 is denoted as R_2.

The information matrix is first expanded by adding zeros for the new feature f_3, then equation (2.3) is used to update the information matrix. The resulting state vector and the information matrix are

$$
\begin{bmatrix} X_{r1} \\ X_{r2} \\ X_{m1} \\ X_{m2} \\ X_{m3} \end{bmatrix}, \quad
\begin{bmatrix}
*_{(3\times3)} & *_{(3\times3)} & *_{(3\times2)} & *_{(3\times2)} & 0_{(3\times2)} \\
*_{(3\times3)} & *_{(3\times3)} & 0_{(3\times2)} & *_{(3\times2)} & *_{(3\times2)} \\
*_{(2\times3)} & 0_{(2\times3)} & *_{(2\times2)} & 0_{(2\times2)} & 0_{(2\times2)} \\
*_{(2\times3)} & *_{(2\times3)} & 0_{(2\times2)} & *_{(2\times2)} & 0_{(2\times2)} \\
0_{(2\times3)} & *_{(2\times3)} & 0_{(2\times2)} & 0_{(2\times2)} & *_{(2\times2)}
\end{bmatrix}
\tag{2.10}
$$

where each $*_{(.)}$ represents a non-zero block with its dimension. Again, it can be noted that non-zero off-diagonal block only appear to the pairs

(X_{r2}, X_{m2}) and (X_{r2}, X_{m3}) in which the states are included in the same observation equation in (2.9) concurrently.

To summarize, after each of the three operations that have been considered in this subsection, non-zero off-diagonal blocks in the information matrix only appear where the two related states are involved concurrently in one of the equations in (2.1), (2.6) and (2.9), as is demonstrated by the information matrices in (2.4), (2.8) and (2.10). It is worth noting that the sparsity pattern of the information matrix in (2.10) corresponds to the graph in Figure 2.1 in the sense that there is a non-zero off-diagonal block only where there is a link in the graph.

Obviously, the above conclusion about the appearance of non-zero off-diagonal blocks holds for the general form of the state vector and any number of observations/control measurements. A typical information matrix from a full SLAM implementation is shown in Figure 2.2, in which the non-zero elements of the information matrix are displayed in black. The sparse structure is obvious.

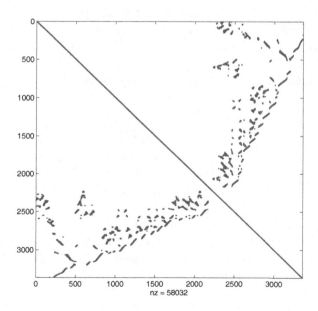

Fig. 2.2   A typical full SLAM information matrix (non-zero elements are displayed in black).

2.2   Information Matrix in the Conventional EIF SLAM Formulation

In the conventional EIF SLAM formulation, all observed features and only the last robot pose are present in the state vector. All the information collected by the estimator can be summarized using the joint posterior over this smaller state vector, as the states for the robot pose constitute Markov sequences [Bar-Shalom *et al.* (2001)] and the final estimate summarizes probabilistically the past. Thus, this SLAM formulation is feasible.

The information matrix evolves in exactly the same manner as in the case of full SLAM when fusing an observation and initializing a new feature. The only difference occurs in the prediction step after the new pose is added, as the past robot pose has to be marginalized out. After each prediction step, there is only one robot pose in the state vector, which is the last one.

In this subsection, the effect of marginalizing out the past robot pose on the information matrix is examined in detail using the example presented in Section 2.1. In order to make the effect more clear without affecting the final result, the analysis starts from the information matrix in (2.10) when the new robot pose X_{r2} has been added into the state vector and the observations made at X_{r2} have been fused. The effect of marginalizing out X_{r1} is analyzed.

Let the information matrix in (2.10) be partitioned as

$$
\left[
\begin{array}{c|cccc}
*_{(3\times3)} & *_{(3\times3)} & *_{(3\times2)} & *_{(3\times2)} & 0_{(3\times2)} \\
\hline
*_{(3\times3)} & *_{(3\times3)} & 0_{(3\times2)} & *_{(3\times2)} & *_{(3\times2)} \\
*_{(2\times3)} & 0_{(2\times3)} & *_{(2\times2)} & 0_{(2\times2)} & 0_{(2\times2)} \\
*_{(2\times3)} & *_{(2\times3)} & 0_{(2\times2)} & *_{(2\times2)} & 0_{(2\times2)} \\
0_{(2\times3)} & *_{(2\times3)} & 0_{(2\times2)} & 0_{(2\times2)} & *_{(2\times2)}
\end{array}
\right]
=
\begin{bmatrix} A & B^T \\ B & C \end{bmatrix}
\tag{2.11}
$$

where the symmetry of the information matrix is utilized.

In order to marginalize out the block A in (2.11), which corresponds to X_{r1}, the Schur complement of A is used [Frese (2005)] and the new information matrix is calculated by

$$
I_{new} = C - BA^{-1}B^T.
\tag{2.12}
$$

Let the matrix B in (2.11) be partitioned as $[B_1^T, B_2^T, B_3^T, 0]^T$. Then the term $BA^{-1}B^T$ in (2.12) can be written as

$$
\begin{bmatrix} B_{1(3\times3)} \\ B_{2(2\times3)} \\ B_{3(2\times3)} \\ 0_{(2\times3)} \end{bmatrix} A_{3\times3}^{-1} \begin{bmatrix} B_{1(3\times3)}^T & B_{2(3\times2)}^T & B_{3(3\times2)}^T & 0_{(3\times2)} \end{bmatrix}
$$

$$
= \begin{bmatrix} B_1 A^{-1} B_{1(3\times3)}^T & B_1 A^{-1} B_{2(3\times2)}^T & B_1 A^{-1} B_{3(3\times2)}^T & 0_{(3\times2)} \\ B_2 A^{-1} B_{1(2\times3)}^T & B_2 A^{-1} B_{2(2\times2)}^T & B_2 A^{-1} B_{3(2\times2)}^T & 0_{(2\times2)} \\ B_3 A^{-1} B_{1(2\times3)}^T & B_3 A^{-1} B_{2(2\times2)}^T & B_3 A^{-1} B_{3(2\times2)}^T & 0_{(2\times2)} \\ 0_{(2\times3)} & 0_{(2\times2)} & 0_{(2\times2)} & 0_{(2\times2)} \end{bmatrix}. \tag{2.13}
$$

Note that the non-zero blocks in (2.13) decide about the new non-zero blocks that are introduced into I_{new} in (2.12) and they are dependent on the structure of B.

After marginalization, the state vector and the information matrix become

$$
\begin{bmatrix} X_{r2} \\ X_{m1} \\ X_{m2} \\ X_{m3} \end{bmatrix}, \quad \begin{bmatrix} *_{(3\times3)} & \star_{(3\times2)} & *_{(3\times2)} & *_{(3\times2)} \\ *_{(2\times3)} & *_{(2\times2)} & \star_{(2\times2)} & 0_{(2\times2)} \\ *_{(2\times3)} & \star_{(2\times2)} & *_{(2\times2)} & 0_{(2\times2)} \\ *_{(2\times3)} & 0_{(2\times2)} & 0_{(2\times2)} & *_{(2\times2)} \end{bmatrix} \tag{2.14}
$$

where each $\star_{(.)}$ represents a non-zero block that has been introduced due to the marginalization and its dimension. The sparsity pattern of the information matrix in (2.14) corresponds to the graph in Figure 2.3. The dashed lines in Figure 2.3 indicate new links caused by the marginalization.

Extent of non-zero off-diagonal blocks that are introduced into the information matrix depends on the structure of the matrix B, which captures the relation between the states which are marginalized out and the remaining states. By examining (2.11) and (2.14), it can be found that non-zero blocks (denoted by $\star$) are introduced into intersections among the states (X_{r2}, X_{m1}, X_{m2}) which have non-zero off-diagonal terms associated with the robot pose (X_{r1}) which is marginalized out. This occurs due to the non-zero blocks at the corresponding location in the matrix B. So after marginalization, the states X_{r2}, X_{m1} and X_{m2} are fully connected. In contrast, in (2.14) the blocks at the intersections between X_{m3} and the other two features, X_{m1} and X_{m2}, remain zero. This is due to the zero off-diagonal block at the intersection between X_{r1} and X_{m3} in (2.11), as is reflected by the zero block at the bottom of the matrix B.

In the conventional EIF SLAM formulation, with the past robot poses being marginalized out constantly, the information matrix gradually be-

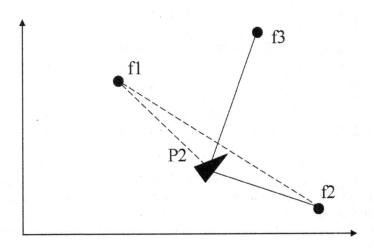

Fig. 2.3   Effect of marginalizing out P1 on the 2D SLAM example.

comes dense. However, the normalized information matrix is found to be approximately sparse in experiments [Thrun *et al.* (2004a)], i.e. many of the off-diagonal elements are very close to zero. The theoretical explanation for this observation is presented in [Frese (2005)]. Figure 2.4 shows a typical normalized information matrix from the conventional EIF SLAM. Non-zero elements of the information matrix are displayed in black. The bigger the number, the darker the point in the figure.

2.3   Meaning of Zero Off-diagonal Elements in Information Matrix

In this subsection, the meaning of zero off-diagonal elements in the general information matrix is examined from a theoretical point of view. The existence of these zero elements in SLAM information matrix is analyzed and then demonstrated using an example.

In the study on Gaussian Markov random fields (GMRF), the canonical representation has been used to facilitate examining the properties of multivariate Gaussian distributions. The sparse structure of the information matrix (also termed as the precision matrix) helps to understand the GMRF through conditional independence. The relationship between zero

Fig. 2.4   A typical normalized information matrix from conventional EIF SLAM (non-zero elements are displayed in black and their values are indicated by the darkness).

off-diagonal elements of the information matrix and conditional indepen-dencies of the associated variables can be characterized [Speed *et al.* (1986)]. Especially, in [Wermuth (1976)], it is shown that an off-diagonal element in the information matrix is zero if and only if the two related variables are conditionally independent given all the other variables.

It is well known that in SLAM conditional independence exists among features if all the robot poses in history are kept in the state vector [Thrun *et al.* (2004b)]. More general analysis can be carried out by using the graph representation of SLAM and by examining the decomposition of the SLAM posterior.

The conditional independence graph can be used to describe the con-ditional independence among the variables involved in a joint Gaus-sian probability density function. The Markov property of the graph indicates that the variables are conditionally independent if the corre-sponding nodes in the graph can be separated [Borgelt *et al.* (2002)]. However, obtaining these graphs for a given probability density distribution is nontrivial. Fortunately, the undirected conditional independence graphs are intimately connected to the decomposition of the probability density function [Lauritzen (1996)]. The combination of these two results in a structure, the well-known Markov network, which provides the qualitative information about the conditional independence statements among the variables in the form of a undirected conditional independence graph [Borgelt *et al.* (2002)].

In the following, the decomposition of the posterior probability density function of the estimate in full SLAM will be analyzed and the corresponding undirected conditional independence graph will be obtained for an example.

In general full SLAM, suppose there are $l+1$ robot poses, $X_{r0}, \ldots, X_{rl}$, and s features, $X_{m1}, \ldots, X_{ms}$, in the state vector. Let the control measurement and observation of each time step be u_k and $z_k (k = 1 \ldots l)$ respectively. With assumed data association, the SLAM posterior can be decomposed as [Thrun *et al.* (2005)]

$$
\begin{aligned}
&p(X_{r(0:l)}, X_{m(1:s)} | z_{1:l}, u_{1:l}) \\
&= cp(X_{r0}) \prod_k p(X_{rk} | X_{r(k-1)}, u_k) p(z_k | X_{rk}, X_{m(1:s)}) \\
&= cp(X_{r0}) \prod_k \left[p(X_{rk} | X_{r(k-1)}, u_k) \prod_i p(z_k^i | X_{rk}, X_{mj_i}) \right]
\end{aligned}
\tag{2.15}
$$

where c is the normalizer and $p(X_{r0})$ is the prior. Note z_k^i is the i-th observation in z_k and with assumed data association X_{mj_i} is the feature involved in this observation. It can be observed that the term $p(X_{rk} | X_{r(k-1)}, u_k)$ incorporates the motion update; and the term $p(z_k | X_{rk}, X_{m(1:s)})$ incorporates the observation update.

With the SLAM posterior decomposed, the undirected conditional independence graph can be obtained accordingly. The graph for the full SLAM example in Section 2.1 is shown in Figure 2.5. Obviously, the nodes f_1, f_2 and f_3 are separated from each other by P_1 and P_2; f_3 and P_1 are separated by P_2; f_1 and P_2 are separated by P_1. These separations indicate conditional independencies among the related states given all the others and correspond to zero off-diagonal blocks in the information matrix in (2.10).

With the presence of conditional independencies among the states in full SLAM, the corresponding information matrix is bound to be exactly sparse. However, in the conventional EIF SLAM formulation, it is known that the marginalization operation destroys conditional independence statements [Paskin (2003)], thus reduces the sparseness of the information matrix. Therefore, how to keep these conditional independencies and thus the

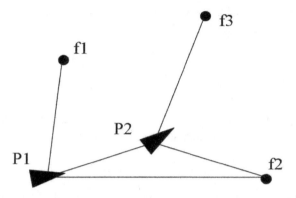

Fig. 2.5   Undirected conditional independence graph of the full SLAM example.

sparseness of the information matrix is a crucial issue for developing exactly sparse information filter algorithms.

2.4   Conditions for Achieving Exact Sparseness

In the full SLAM problem, where the state vector includes all feature locations and all robot poses from which the observations are made, the mechanism of the EIF results in an exactly sparse information matrix [Frese (2005)]. As discussed in Section 2.1, when fusing new information using the EIF, only parts of the information matrix that are directly influenced by the new information need to be updated or initialized. For example, during a prediction step, only the diagonal and off-diagonal elements relating the two robot poses that are related by the process model are affected; during an update step, only the diagonal and off-diagonal elements relating the robot pose and the features that are being observed are affected. Numerical values of all other elements in the information matrix remain unchanged. Therefore, one condition for obtaining an exactly sparse information matrix is that each observation relates to only a fraction of the states in the state vector and that each state is only involved in observations simultaneously with a subset of the states in the state vector.

It is the operation of marginalizing out a part of the state that causes fill-ins in the information matrix and makes it dense [Eustice *et al.* (2006)]. As discussed in Section 2.2, this is because in the EIF the marginalization operation involves a subtraction operation with a component in the Schur complement [Frese (2005)]. As a general example, suppose the original information matrix I is sparse and is of the form

$$I = \begin{bmatrix} A & B^T \\ B & C \end{bmatrix} \tag{2.16}$$

where A, B, C are matrices with the dimensions omitted and A is invertible. A specific example of these matrices is available in (2.11). If the states corresponding to the block matrix A are to be marginalized out, the new information matrix I_{new}, which is of the same dimension as C, can be computed as

$$I_{new} = C - BA^{-1}B^T \tag{2.17}$$

in which the matrix, A^{-1}, is normally dense. The result of the marginalization operation is that the new information matrix I_{new} becomes dense even if the original information matrix I is sparse. The structure of the matrix B, which records the connections by non-zero off-diagonal blocks between the states to be marginalizing out (corresponding to A) and the remaining states (corresponding to C), controls the extent of fill-ins.

If the matrix B is made close to a null matrix, the extent of fill-ins caused by the Schur complement can be reduced. Therefore, another condition for obtaining an exactly sparse information matrix is that the operation of marginalizing out past robot poses is eliminated or its effect is reduced, for example, by controlling the structure of the matrix B in above description.

To sum up, the conditions for obtaining an exactly sparse information matrix in SLAM are:

1) Each observation relates to only a fraction of states in the state vector. It is also required that each state is only involved in observations simultaneously with a subset of all the states in the state vector no matter how many observations are made. This occurs naturally in SLAM when the sensor range is limited, as two features will not be simultaneously observed if the distance between them is more than twice of the sensor range.

2) The operation of marginalizing out past robot poses which are linked by odometry does not exist in the filter steps or its effect is reduced. This can be done, for example, by controlling the connections by non-zero off-diagonal blocks between the robot pose which is to be marginalized out and the other states in the state vector.

These two conditions are closely related to the conditional independence analysis in Section 2.3. When a feature is observed at a robot pose, the robot pose and the feature appear in the same observation equation, indicating direct dependence among the corresponding states. In other words, these states are not conditionally independent. The first condition puts a limit on the number of dependence statements in this scenario. As the discussion in Section 2.3 manifests, marginalization of robot poses causes loss of conditional independence among the states. The second condition is to reduce this effect.

The above two conditions are sufficient to achieve an exactly sparse information matrix in SLAM. The extent of sparseness of the information matrix depends on how many states are involved in each observation simultaneously, and how the effect of marginalization is controlled, if it exists.

2.5 Strategies for Achieving Exact Sparseness

This section classifies the exactly sparse information filter based SLAM algorithms in this book, based on the strategies used to achieve the sparseness according to the two conditions discussed in Section 2.4. For the details of other valuable contributions that use approximately or exactly sparse representations such as [Paskin (2003); Thrun *et al.* (2004a); Folkesson and Christensen (2004); Thrun and Montemerlo (2004); Frese *et al.* (2005); Frese (2006b); Dellaert and Kaess (2006); Eustice *et al.* (2006); Walter *et al.* (2007)], the readers are referred to the relevant papers.

2.5.1 *Decoupling Localization and Mapping*

If the robot pose is not in the state vector, then the marginalization operation does not exist in the filter process. According to the conditions discussed in Section 2.4, the associated information matrix in this case can be made exactly sparse. D-SLAM (Chapter 3) uses a state vector that only contains the feature locations to generate maps of an environment. Robot location estimate is obtained through a separate, concurrent process.

In D-SLAM mapping, the original measurements relating the robot and features are first transformed into relative distances and angles among features. Then these transformed measurements are fused into the map using EIF. It is shown that only the features that are observed at the same time instant have links (non-zero off-diagonal blocks) in the information matrix. The marginalization operation obviously does not exist. Therefore, the information matrix is exactly sparse.

The fact that the process information is not exploited in building the map leads to some information loss.

2.5.2   *Using Local Submaps*

By using submaps, the effect of the marginalization operation on the information matrix can be restricted in each individual submap. On the global level, the information matrix can be made exactly sparse if the relative geometrical location among the submaps is resolved and the size of the submaps is controlled properly. The Sparse Local Submap Joining Filter (SLSJF) (Chapter 5) is a local map joining algorithm for mapping large scale environments based on this idea.

In SLSJF, the local maps, treated as virtual observations made from the robot start pose of each local map, are fused by using an EIF to form a global representation of the environment. On the global level, the first condition in Section 2.4 is satisfied as long as the size of the local maps is small compared with the global map. It should be noted that when fusing the local map, the virtual observation, into the global map, all the robot poses and feature locations that are present in this local map are linked in the global information matrix because they are correlated in the local map. In other words, they appear in the same observation equation.

Most of the operations to marginalize out old robot poses are incorporated in building the local map, thus their effect is restricted within the local map and they do not directly affect the sparseness of the global information matrix. The only exceptions are the ones between two successive local maps. By keeping the start and end robot poses of each local map into the global state vector, these marginalization operations are eliminated.

It is worth noting that the global information matrix in SLSJF is made exactly sparse without any information loss. The only expense is that the global state vector is expanded by adding the start and end robot poses of each local map, but the number of these poses is much smaller than the number of features.

2.5.3   *Combining Decoupling and Submaps*

Obviously, the combination of the decoupling idea and submaps also leads to exact sparseness, which has advantages in mapping large scale environments. The D-SLAM Local Map Joining Filter (Chapter 4) uses this idea. It is a submap based algorithm to map large scale environments using the D-SLAM framework. Only feature locations are kept in the global state vector and the information matrix of the global map is exactly sparse.

Two steps are taken to fuse the local map. First relative position information among the feature in the local map is extracted. Then the D-SLAM framework is used to fuse the relative information and build a global map. As discussed in Section 2.5.2, the first condition presented in Section 2.4 is satisfied as long as the size of the local map is small.

As discussed in Section 2.5.2, the operations to marginalize out past robot poses in local maps do not directly affect the sparseness of the global information matrix. The marginalization operations between two successive local maps is not applied in this algorithm because no robot pose is included in the global state vector.

As the robot pose is removed when fusing the local map into the global map, there is some information loss in this algorithm.

2.6 Important Practical Issues in EIF SLAM

A number of important practical issues need to be dealt with when using sparse information filters for SLAM.

In particular, the recovery of the relevant elements in the state vector is required to compute the Jacobians which are needed for the linearization of the observation and process models. Probabilistic data association requires the recovery of the relevant elements in the state vector as well as the associated covariances. Although the filter steps are made efficient due to the presence of the exactly sparse information matrix, the need for state and covariance recovery may have a significant impact on the computational cost of a practical implementation.

Data association is also a computationally costly operation. It is easier when other information such as feature descriptors extracted using the Scale Invariant Feature Transform (SIFT) [Lowe (2004)] in vision based SLAM is available. When this is not feasible, probabilistic strategies for associating measurements with the features in the map are required each time a feature is observed. Although the computational cost can be somewhat mitigated by identifying a subset of features from which the observation possibly originate, data association (including the recovery of the relevant elements in the state vector and the associated covariances) becomes arguably the most expensive operation in all EIF-based SLAM algorithms.

Information loss is one of the tradeoffs that need to be considered when computing strategies are used to obtain an exactly sparse information matrix. Manipulating the way observation information is exploited, for example recasting the measurements as proposed in D-SLAM, lead to a suboptimal estimate. Manipulating the way the process model is used, for example by "kidnapping" and "relocating" the robot as shown in the Exactly Sparse Extended Information Filter (ESEIF) [Walter *et al.* (2007)], also causes information loss. It is worth noting that including all robot poses from which the observations are made into the state vector does not improve the information content of the map or the final robot pose due to the Markov property of the states for the robot pose [Bar-Shalom *et al.* (2001)]. When the effect due to linearization errors is not considered, the EIF SLAM formulation where the state vector contains the final robot pose and all the feature locations is optimal.

2.7   Summary

In this chapter, the evolution of the information matrix under the filter operations in the full SLAM and conventional EIF SLAM formulations are analyzed in detail. The connection between the zero off-diagonal elements in the information matrix and the conditional independence statements of the states are manifested. Then the existence of these conditional independence statements in SLAM is analyzed. Two conditions that need to be satisfied to achieve exact sparseness of the information matrix in the filter process are described and the strategies that have been exploited by the algorithms in this book are examined according to these two conditions.

In EIF-based SLAM algorithms, the state vector and covariance matrix are not maintained explicitly. The recovery of these is needed to linearize the nonlinear process and observation models and perform stochastic data association. The recovery process can require significant computational effort and data association itself is a computationally costly operation. Also, manipulating the filter steps to achieve the exactly sparse information matrix may lead to information loss. These practical issues need to be considered in the development of EIF-based SLAM algorithms.

Chapter 3

Decoupling Localization and Mapping

This chapter presents a SLAM algorithm, in which the SLAM problem is reformulated such that the mapping and localization are treated as two concurrent yet separated processes [Wang *et al.* (2007a)]. The new formulation is achieved by transforming the measurement vector into two parts, one containing information relating features in the map and the other containing information relating the map and the robot. With this framework, SLAM with a range and bearing sensor in an environment populated with point features can be decoupled into solving a nonlinear static estimation problem for mapping and a low-dimensional dynamic estimation problem for localization. The decoupled structure of the filter leads to the name of the algorithm, D-SLAM.

As a consequence of decoupling localization and mapping in SLAM, the information matrix of D-SLAM mapping becomes exactly sparse. This is because the robot prediction step that includes the marginalization operation and that makes the information matrix dense is not required during mapping. A significant saving in computational cost can be achieved for large scale problems by exploiting the special properties of sparse matrices, when an EIF is used in the estimation process.

Other contributions of this chapter include two methods to efficiently recover the feature location estimates and the associated covariances from the information vector and the sparse information matrix. One method uses Preconditioned Conjugated Gradient (PCG) and the other uses complete Cholesky Factorization. In the method using PCG, a preconditioner is calculated by factorizing the information matrix using an incremental approximate Cholesky Factorization procedure, so that the Conjugated Gradient algorithm terminates in a few iterations. In the method using complete Cholesky Factorization, the factorization of the information matrix is performed using an incremental procedure. With either of these two methods, the feature location estimates and the exact covariances can be recovered without any approximation.

This chapter is organized as follows. In Section 3.1, the main steps of D-SLAM are stated, including measurement transformation, mapping and localization. Section 3.2 discusses why the information matrix of D-SLAM mapping is exactly sparse. In Section 3.3, two efficient state and covariance recovery methods are introduced. Section 3.4 addresses implementation issues in D-SLAM such as admissible measurements and data association. Simulation results are presented in Section 3.5 and experimental results using indoor and outdoor practical data are presented in Section 3.6 to evaluate the algorithm. In Section 3.7, the computational complexity of D-SLAM is analyzed in detail. Section 3.8 discusses the consistency of D-SLAM. In Section 3.9, the work in the literature on decoupling SLAM and achieving sparseness in EIF SLAM is discussed. The chapter is summarized in Section 3.10.

3.1   The D-SLAM Algorithm

3.1.1   *Extracting Map Information from Observations*

Observations made from a sensor mounted on the robot contain the relative location of the features with respect to the robot as shown in Figure 3.1. Using these observations directly in SLAM makes the robot pose and feature location estimates correlated. In order to decouple mapping and localization in SLAM, a key step is to extract map information from the observations. This can be achieved by transforming the measurement vector into one consisting of two parts: one part containing distances and angles among features; the other relating the features and the robot location. It is important to formulate the information extraction process in order to minimize information loss to maintain efficiency, and to avoid information reuse.

To maintain notational simplicity and improve clarity, it is assumed that all observations contain measurements to at least two features that are previously seen and already present in the map. A strategy to overcome this limitation by combining a number of measurements into one admissible measurement is described in Section 3.4.1.

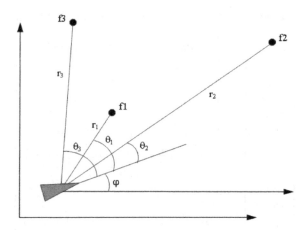

Fig. 3.1   Original range and bearing measurements.

3.1.1.1   *The Original Measurements*

Suppose the robot observes m features $f_1, \cdots, f_m$ at a particular time, among which f_1 and f_2 have been previously seen.

As shown in Figure 3.1, the original measurements provide the relative location of the features with respect to the robot location. For example, the measurements of feature f_2 are the relative range and bearing (r_2 and θ_2) of f_2 with respect to the robot location. The original measurement vector contains the measured range and bearing of each observed feature:

$$z_{old} = \left[r_1, \theta_1, \cdots, r_m, \theta_m \right]^T, \tag{3.1}$$

which contains noise, assumed to be white Gaussian with zero mean and a covariance matrix

$$R_{old} = diag \left[\sigma_{r_1}^2, \sigma_{\theta_1}^2, \cdots, \sigma_{r_m}^2, \sigma_{\theta_m}^2 \right]. \tag{3.2}$$

3.1.1.2   *Distances and Angles with respect to f_1 and f_2*

The original measurements relate the robot pose and the feature locations. They need to be transformed into relative quantities relating only feature locations. Obviously, distances between two features and angles among three features are relative quantities which are convenient to use.

The original measurement vector can be transformed to z_{new} written as

$$
\begin{bmatrix} \alpha_{r12} \\ d_{1r} \\ \alpha_{\phi 12} \\ -\,-\,- \\ d_{12} \\ \alpha_{312} \\ d_{13} \\ \vdots \\ \alpha_{m12} \\ d_{1m} \end{bmatrix} =
\begin{bmatrix}
atan2\left(\frac{-\tilde{y}_1}{-\tilde{x}_1}\right) - atan2\left(\frac{\tilde{y}_2-\tilde{y}_1}{\tilde{x}_2-\tilde{x}_1}\right) \\
\sqrt{(-\tilde{x}_1)^2 + (-\tilde{y}_1)^2} \\
-atan2\left(\frac{\tilde{y}_2-\tilde{y}_1}{\tilde{x}_2-\tilde{x}_1}\right) \\
-\,-\,- \\
\sqrt{(\tilde{x}_2 - \tilde{x}_1)^2 + (\tilde{y}_2 - \tilde{y}_1)^2} \\
atan2\left(\frac{\tilde{y}_3-\tilde{y}_1}{\tilde{x}_3-\tilde{x}_1}\right) - atan2\left(\frac{\tilde{y}_2-\tilde{y}_1}{\tilde{x}_2-\tilde{x}_1}\right) \\
\sqrt{(\tilde{x}_3 - \tilde{x}_1)^2 + (\tilde{y}_3 - \tilde{y}_1)^2} \\
\vdots \\
atan2\left(\frac{\tilde{y}_m-\tilde{y}_1}{\tilde{x}_m-\tilde{x}_1}\right) - atan2\left(\frac{\tilde{y}_2-\tilde{y}_1}{\tilde{x}_2-\tilde{x}_1}\right) \\
\sqrt{(\tilde{x}_m - \tilde{x}_1)^2 + (\tilde{y}_m - \tilde{y}_1)^2}
\end{bmatrix} \tag{3.3}
$$

where

$$\tilde{x}_i = r_i \cos\theta_i, \quad \tilde{y}_i = r_i \sin\theta_i, \quad i = 1, \cdots, m. \tag{3.4}$$

Essentially, the quantities in z_{new} are obtained by recalculating the the robot pose and the feature locations with respect to the selected reference point, f_1. The physical meaning of the new measurement vector is shown in Figure 3.2, in which $d_{12}, \alpha_{312}, d_{13}$ are distances and angles among features. In Figure 3.2, the geometric interpretation is demonstrated for the measurements to f_1, f_2 and f_3. Extending this to the measurements to $f_4, \cdots, f_m$ is straightforward. These relative quantities contain information about the map only, and are used in D-SLAM mapping.

It should be noted that f_1 and f_2 are chosen to be the reference points only to facilitate the explanation. In fact, the reference points can be any two features, which have been previously observed, in the current measurements.

The last $2m - 3$ elements in the measurement vector shown in (3.3) contain information about distances and angles among features that are independent of the coordinate frame. The first three elements depend on the robot pose and the locations of features f_1, f_2. This part carries information about the robot location. Thus the measurement vector can be naturally partitioned into two vectors denoted as

$$z_{new} = \begin{bmatrix} z_{rob} \\ z_{map} \end{bmatrix} \tag{3.5}$$

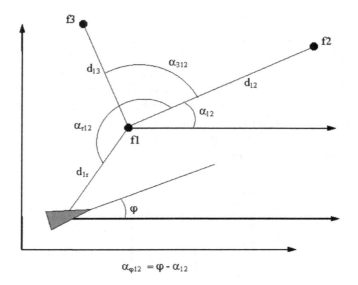

$$\alpha_{\varphi 12} = \varphi - \alpha_{12}$$

Fig. 3.2 Geometric interpretation of the measurement vector given by equation (3.3) for features f_1, f_2 and f_3.

where

$$z_{rob} = \begin{bmatrix} \alpha_{r12} \\ d_{1r} \\ \alpha_{\phi 12} \end{bmatrix}, \quad z_{map} = \begin{bmatrix} d_{12} \\ \alpha_{312} \\ d_{13} \\ \vdots \\ \alpha_{m12} \\ d_{1m} \end{bmatrix}. \tag{3.6}$$

There is a one to one correspondence between z_{old} and z_{new}. Furthermore, z_{new} is constructed such that it contains the minimum number of elements required to completely capture the total information content contained in z_{old}. The elements in z_{new} are sufficient and necessary to express z_{old}. In other words, all elements in z_{new} are needed to express z_{old} (necessary condition) and they are sufficient to express z_{old} (sufficient condition).

It is important, however, to note that the two measurement vectors z_{rob} and z_{map} are not independent. Thus the proposed transformation does not completely divide the information contained in the original measurements into two parts. Therefore, the estimation process that exploits these new measurement vectors needs to be structured appropriately in order to avoid statistical inconsistency.

3.1.1.3 *Measurement Noise Covariances*

For a Gaussian random (vector) variable x with a mean $\bar{x}$ and a covariance matrix R_x, any (vector) function of x, $g(x)$, can be approximated by a Gaussian provided x is near $\bar{x}$. The mean of this Gaussian is $g(\bar{x})$ and its covariance matrix is $\nabla g R_x \nabla g^T$ where ∇g is the Jacobian of g with respect to x evaluated at $\bar{x}$. This relationship can be used to compute the covariance matrix of the noise on the new measurement vector z_{map} from (3.2), (3.3), (3.4) and (3.6) as follows.

The relation in (3.4) can be written as

$$\tilde{z}_{old} = \begin{bmatrix} \tilde{x}_1 \\ \tilde{y}_1 \\ \vdots \\ \tilde{x}_m \\ \tilde{y}_m \end{bmatrix} = \begin{bmatrix} r_1 cos\theta_1 \\ r_1 sin\theta_1 \\ \vdots \\ r_m cos\theta_m \\ r_m sin\theta_m \end{bmatrix} = T_1(z_{old}). \tag{3.7}$$

From (3.3), (3.4) and (3.6)

$$z_{map} = \begin{bmatrix} \sqrt{(\tilde{x}_2 - \tilde{x}_1)^2 + (\tilde{y}_2 - \tilde{y}_1)^2} \\ atan2\left(\frac{\tilde{y}_3 - \tilde{y}_1}{\tilde{x}_3 - \tilde{x}_1}\right) - atan2\left(\frac{\tilde{y}_2 - \tilde{y}_1}{\tilde{x}_2 - \tilde{x}_1}\right) \\ \sqrt{(\tilde{x}_3 - \tilde{x}_1)^2 + (\tilde{y}_3 - \tilde{y}_1)^2} \\ \vdots \\ atan2\left(\frac{\tilde{y}_m - \tilde{y}_1}{\tilde{x}_m - \tilde{x}_1}\right) - atan2\left(\frac{\tilde{y}_2 - \tilde{y}_1}{\tilde{x}_2 - \tilde{x}_1}\right) \\ \sqrt{(\tilde{x}_m - \tilde{x}_1)^2 + (\tilde{y}_m - \tilde{y}_1)^2} \end{bmatrix} = T_2(\tilde{z}_{old}). \tag{3.8}$$

The covariance matrix of z_{map} is given by

$$R_{map} = \nabla T_2 \nabla T_1 R_{old} \nabla T_1^T \nabla T_2^T \tag{3.9}$$

where ∇T_1 is the Jacobian of the function T_1 with respect to all the elements in z_{old} evaluated on the current observation z_{old}, ∇T_2 is the Jacobian of the function T_2 with respect to all the elements in $\tilde{z}_{old}$ evaluated on the current value of $\tilde{z}_{old}$, and R_{old} is given in (3.2).

3.1.2 *Key Idea of D-SLAM*

In D-SLAM, the key idea is to use z_{map} to estimate a state vector containing only the locations of features in the environment. As shown in the following sections, the absence of the robot location from the state vector results in an estimator structure that offers significant computational advantages.

As z_{rob} and z_{map} are not independent, z_{rob} also contains some information about the map which is not exploited during the mapping process, resulting in some information loss. Furthermore, this dependency makes it important that the localization process is formulated carefully in order to avoid information reuse. The details of the mapping and localization algorithms in D-SLAM are described in the following subsections.

3.1.3 *Mapping*

3.1.3.1 *State Vector*

The state vector used for mapping contains the locations of the features:

$$X = (X_1, \cdots, X_n)^T = (x_1, y_1, x_2, y_2, \cdots, x_n, y_n)^T, \qquad (3.10)$$

where $X_1, \cdots, X_n$ denote the locations of features $f_1, \cdots, f_n$ respectively.

For convenience, the initial robot pose is used to define the coordinate frame, where the origin is the initial robot position and the x-axis is along the initial robot heading. The information vector $i(k)$ is defined as

$$i(k) = I(k)\hat{X}(k) \qquad (3.11)$$

where $I(k)$ is the information matrix which is the inverse of the state covariance matrix $P(k)$ and $\hat{X}(k)$ is the estimate of the state. As the features are stationary, mapping reduces to a static nonlinear estimation problem which can be formulated in the information form and performed efficiently using an EIF (e.g. [Maybeck (1979); Thrun *et al.* (2004a)]) as follows.

3.1.3.2 *Measurement Model*

Suppose the robot observes m features $f_1, \cdots, f_m$, in which two features f_1, f_2 have been previously seen. The recast measurement used for mapping is

$$\begin{aligned} z_{map} &= [d_{12}, \alpha_{312}, d_{13}, \cdots, \alpha_{m12}, d_{1m}]^T \\ &= H_{map}(X) + w_{map} \end{aligned} \qquad (3.12)$$

where

$$H_{map}(X) = \begin{pmatrix} \sqrt{(x_2 - x_1)^2 + (y_2 - y_1)^2} \\ atan2\left(\frac{y_3 - y_1}{x_3 - x_1}\right) - atan2\left(\frac{y_2 - y_1}{x_2 - x_1}\right) \\ \sqrt{(x_3 - x_1)^2 + (y_3 - y_1)^2} \\ \cdots \\ atan2\left(\frac{y_m - y_1}{x_m - x_1}\right) - atan2\left(\frac{y_2 - y_1}{x_2 - x_1}\right) \\ \sqrt{(x_m - x_1)^2 + (y_m - y_1)^2} \end{pmatrix} \qquad (3.13)$$

and w_{map} is the zero mean Gaussian measurement noise whose covariance matrix R_{map} is given in (3.9).

3.1.3.3 *Feature Initialization*

Current location estimates of the previously seen features f_1, f_2, $\hat{X}_1 = (\hat{x}_1, \hat{y}_1)$ and $\hat{X}_2 = (\hat{x}_2, \hat{y}_2)$, can be used together with d_{1i}, α_{i12} in z_{map} to compute an estimate of the initial location of a new feature f_i as follows:

$$\begin{aligned} \alpha_{12} &= atan2(\tfrac{\hat{y}_2 - \hat{y}_1}{\hat{x}_2 - \hat{x}_1}) \\ \hat{x}_i &= \hat{x}_1 + d_{1i}\cos(\alpha_{12} + \alpha_{i12}) \\ \hat{y}_i &= \hat{y}_1 + d_{1i}\sin(\alpha_{12} + \alpha_{i12}). \end{aligned} \qquad (3.14)$$

These values are needed to linearize the nonlinear measurement model when updating the map. As part of the initialization process, the dimensions of the information vector and the information matrix are also increased by adding an appropriate number of zeros. These zeros in the information vector and the information matrix reflect the fact that information contained in the measurement to f_i is yet to be incorporated [Bar-Shalom *et al.* (2001)].

3.1.3.4 *Map Update*

The information vector and the information matrix can now be updated using the measurement z_{map} by the EIF update equations [Thrun *et al.* (2004a)]:

$$\begin{aligned} I(k+1) &= I(k) + \nabla H_{map}^T R_{map}^{-1} \nabla H_{map} \\ i(k+1) &= i(k) + \nabla H_{map}^T R_{map}^{-1}[z_{map}(k+1) - \\ &\quad H_{map}(\hat{X}(k)) + \nabla H_{map}\hat{X}(k)] \end{aligned} \qquad (3.15)$$

where ∇H_{map} is the Jacobian of the function H_{map} with respect to all the states evaluated on the current state estimate $\hat{X}(k)$.

The estimate $\hat{X}(k+1)$ and the association covariance matrix $P(k+1)$ can be computed using

$$\hat{X}(k+1) = I(k+1)^{-1}i(k+1)$$
$$P(k+1) = I(k+1)^{-1}. \tag{3.16}$$

However, computing the inverse of the information matrix will be computationally expensive except for very small maps. Efficient strategies for recovering the state estimate and part of the covariance matrix will be discussed in Section 3.3.

3.1.4 *Localization*

Given the current map, the robot location can be easily obtained by solving a "kidnapped robot" problem, which is to recover the robot pose using the current map and measurements to some features in the map without any prior knowledge of the robot pose. Suppose among the features that the robot observes, f_1 and f_2 have been previously seen in the current map, as shown in Figure 3.3. The robot pose can be computed using the estimates of the locations of f_1 and f_2 together with the observations made to these features. This solution, however, discards all knowledge of the previous robot location.

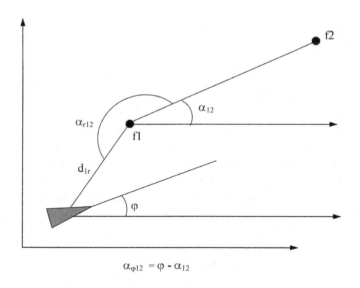

Fig. 3.3 Solving the kidnapped robot problem.

On the other hand, localization can also be done efficiently by an EKF-based local SLAM process where only the features in the vicinity of the robot are retained in the state vector. Features are removed from the state vector once they are not visible from the robot. This is effectively SLAM-aided dead reckoning which provides much better robot location estimate than that obtained using dead reckoning alone. However, the localization results will be much inferior to that obtained from a complete SLAM process as the information contained in the features removed from the state vector is unrecoverable and thus the robot location estimate will not improve during loop closures.

Clearly combining the results from these two approaches can lead to an improved estimate of the robot location. As the two robot location estimates are correlated and the extent of this correlation is unknown, it is proposed that Covariance Intersection (CI) [Chen *et al.* (2002); Julier and Uhlmann (2001a)] is used for this process. Figure 3.4 shows a flow chart illustrating the localization process in D-SLAM. The following subsections describe the localization process in more detail.

3.1.4.1   *State Vector*

The dimension and content of the state vector are different during the seven (A-F) stages in Figure 3.4.

At steps A and B, the state vector contains the locations of the robot and all the features observed at time k. Suppose the robot observes features $f_1, \cdots, f_m$ at time $k+1$, among which $f_1, \cdots, f_{m_1}, m_1 \leq m$ are features that have been previously seen and are already included in Map(k) (step F). The state vector at steps C, G and D contains the locations of the robot and the features observed at time $k+1$ which are in Map(k), $(X_r(k+1), X_1, \cdots, X_{m_1})$. The state vector at step E contains the locations of the robot and all the features observed at time $k+1$, $(X_r(k+1), X_1, \cdots, X_m)$.

In the EKF-based local SLAM process (steps A-B-C-D in Figure 3.4), the content and dimension of the state vector will change often. New elements will be added into the state vector when the robot observes new features, and some elements will be deleted from the state vector when the corresponding features are out of the robot's sight.

The content and dimension of the state vector in the solution to the "kidnapped robot" problem (steps F-G-D in Figure 3.4) are obvious, as they are only governed by the observation to features that already exist in the D-SLAM map at each time step.

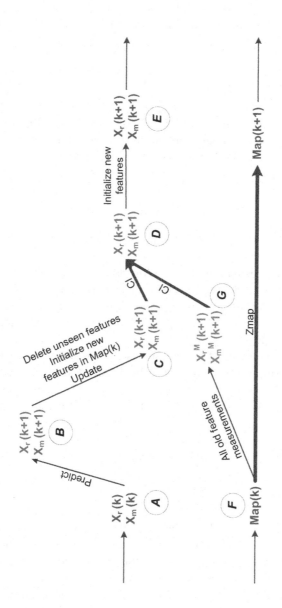

Fig. 3.4   Flow chart of the localization and the mapping processes in D-SLAM.

It is important to note that at steps C and G, the states obtained from the two localization processes should contain the same features and these features should be located in the same order in the two state vectors, so that they can be fused using CI.

3.1.4.2 *Robot Location Estimate 1: Solution to the "Kidnapped Robot" Problem*

The process from F to G in Figure 3.4 is to solve the "kidnapped robot" problem which can be formulated in the information form and solved as a linearized minimum mean square error estimation problem.

The observations to features $f_1, \cdots, f_{m_1}, m_1 \le m$ that have been previously seen are given by

$$
\begin{aligned}
z_{loc} &= (r_1, \theta_1, \cdots, r_{m_1}, \theta_{m_1})^T \\
&= H_{loc}(X_r(k+1), X_1, \cdots, X_{m_1}) + w_{loc},
\end{aligned}
\tag{3.17}
$$

where w_{loc} is the corresponding measurement noise. The state vector used to solve the "kidnapped robot" problem is $[X_r(k+1), X_1, \cdots, X_{m_1}]^T$. The function H_{loc} relates these states and the observations. The estimates of the locations of features $f_1, \cdots, f_{m_1}, m_1 \le m$ are extracted from current D-SLAM map estimate.

To "initialize" the value of the robot pose in the state vector, the robot location is first computed as illustrated by Figure 3.3. Let the estimates of f_1 and f_2 be $\hat{X}_1 = (\hat{x}_1, \hat{y}_1)$ and $\hat{X}_2 = (\hat{x}_2, \hat{y}_2)$. The initial value of the robot pose $\hat{X}_r(k+1) = [\hat{x}_r, \hat{y}_r, \hat{\phi}_r]^T$ can be computed using

$$
\begin{aligned}
\hat{x}_r &= \hat{x}_1 + d_{1r}\cos(atan2(\tfrac{\hat{y}_2-\hat{y}_1}{\hat{x}_2-\hat{x}_1}) + \alpha_{r12}) \\
\hat{y}_r &= \hat{y}_1 + d_{1r}\sin(atan2(\tfrac{\hat{y}_2-\hat{y}_1}{\hat{x}_2-\hat{x}_1}) + \alpha_{r12}) \\
\hat{\phi}_r &= atan2(\tfrac{\hat{y}_2-\hat{y}_1}{\hat{x}_2-\hat{x}_1}) + \alpha_{\phi12}
\end{aligned}
\tag{3.18}
$$

where $z_{rob} = [\alpha_{r12}, d_{1r}, \alpha_{\phi12}]^T$ can be computed from (3.3), (3.4) and (3.6). Then the dimensions of the information vector and the information matrix are increased by adding an appropriate number of zeros. After that, the state containing the robot pose and locations of features $f_1, \cdots, f_{m_1}, m_1 \le m$ is updated using z_{loc} defined in (3.17). This process is similar to that presented in Section 3.1.3.3 and 3.1.3.4 used for initialization and update of features.

3.1.4.3 *Robot Location Estimate 2: SLAM with Local Features*

In this EKF-based local SLAM process, at each time step the state vector only contains the locations of the robot and the features that the robot can currently observe. For example, at time k, the robot observes features $f_{k_1}, \cdots, f_{k_m}$. Then only the locations of these m features are in the state vector, while the locations of all other features that the robot has previously observed are removed. Corresponding rows and columns of the covariance matrix are also removed accordingly. Once the state vector and the covariance matrix are modified in this manner, standard procedures in traditional EKF SLAM for prediction, update and initialization are used in a manner similar to those presented in [Dissanayake *et al.* (2001)].

In Figure 3.4, from step A to step B is the prediction step based on the robot process model

$$X_r(k+1) = f(X_r(k), u(k)) + w(k), \tag{3.19}$$

where $u(k)$ is the control, $w(k)$ is the process noise with zero mean and a covariance Q, and $X_r(k)$ is the robot pose in the state vector at time k, $X(k)$. The dimension of the whole state vector remains unchanged during this step.

Between steps B and C, three operations are performed. First the features which are not observed by the robot at time $k+1$ but still exist in the current state vector are removed. Then the "new features" that have been previously deleted by the local SLAM process and that are reobserved at time $k+1$ are initialized. These features are new to the local SLAM, but already exist in the D-SLAM map. However, the information contained in the estimates of these features in the D-SLAM mapping process is not exploited in order to maintain the consistency of the local SLAM process. Then the state estimate is updated using observations from features maintained by the local SLAM process.

From steps D to E, the new features observed at time $k+1$ are initialized. Note these features have never been observed by the robot, and thus are different from the features initialized between steps B and C.

The loop closing effect, in which estimates of features in the state are improved due to observing previously seen and well defined features will not happen in the local SLAM process. But the information from the process model and the observations to features close to the robot is fused into the state estimate.

3.1.4.4   *Fusing the Two Robot Location Estimates*

The local SLAM estimate is optimal, until the robot closes a loop by re-visiting a previously traversed region of the map and reobserves features that have been removed from the state vector. The "kidnapped robot" solution will be superior when loop closures are present. These two estimates make use of the information from the same measurements, and thus are correlated. Maintaining the correlations is possible but requires significant computational effort. Covariance intersection (CI) [Chen *et al.* (2002); Julier and Uhlmann (2001a)], a technique that facilitates combining two correlated pieces of information, when the extent of correlation itself is unknown is, therefore, used to fuse these two estimates. The criterion used in computing the CI solution is selected to be minimizing the trace of the submatrix associated with the robot pose in the covariance matrix. While other criteria, such as determinant and entropy, can also be used to evaluate the quality of the estimates, trace is selected as minimizing the uncertainty of the robot pose estimate is arguably the most important consideration. It should be noted that the units of the robot position and orientation are different, therefore they need to be appropriately scaled when computing the trace. Scale factor of 1 is adequate if the units of the robot position and orientation are meter and radian respectively.

Suppose X_1 and X_2 are the two estimates that result from the local SLAM process and the "kidnapped robot" solution respectively. Let P_1 and P_2 be the corresponding covariance matrices. A conservative combination of these two estimates with no knowledge of their correlation can be computed using the CI algorithm as follows:

$$P_{CI}^{-1} = \omega P_1^{-1} + (1 - \omega) P_2^{-1}$$
$$P_{CI}^{-1} X_{CI} = \omega P_1^{-1} X_1 + (1 - \omega) P_2^{-1} X_2 \tag{3.20}$$

which leads to

$$P_{CI} = (\omega P_1^{-1} + (1 - \omega) P_2^{-1})^{-1}$$
$$X_{CI} = P_{CI} (\omega P_1^{-1} X_1 + (1 - \omega) P_2^{-1} X_2) \tag{3.21}$$

where X_{CI} and P_{CI} are the resulting state estimate and the associated covariance matrix respectively. The parameter $\omega (0 \leq \omega \leq 1)$ is chosen to minimize the trace of the submatrix associated with the robot pose in the resulting covariance matrix.

This process is illustrated by the steps C, G and D in Figure 3.4. The state vectors at steps C and G should contain the same features in the same order. Thus the state vector of the local SLAM process needs to be reordered between steps B and C. The result of CI (step D in Figure 3.4) will be used in the local SLAM process in the next time step.

In the case when loop closures do not exist, the CI result is identical to the estimate from the local SLAM as no additional information is available from the "kidnapped robot" solution.

It is worth noting that the "kidnapped robot" solution carries information from the mapping component of D-SLAM, and that this information will flow through to the estimate of the state vector used by the local SLAM. As the errors in the D-SLAM map will be bounded, provided that loop closures exist, the errors in the robot pose estimate from the local SLAM process will stay bounded.

3.2 Structure of the Information Matrix in D-SLAM

Using a typical sensor with the sensor range limited, only the features that are in close proximity will be simultaneously observed. Thus, if feature f_i and feature f_j are far away from each other, the measurement z_{map} will never contain both f_i and f_j. As the information matrix update involves a simple addition operation, the off-diagonal elements relating to f_i and f_j in the information matrix will remain exactly zero. Therefore, for a large scale map, a significant portion of the information matrix will be exactly zero, resulting in an exactly sparse information matrix.

When the robot observes $f_{k_1}, \cdots, f_{k_m}$, without loss of generality, suppose that these m features are at the bottom of the state vector. The Jacobian ∇H_{map} in (3.15) will be

$$\nabla H_{map} = \left[0, \cdots, 0, \frac{\partial H_{map}}{\partial X_{k_1}}, \cdots, \frac{\partial H_{map}}{\partial X_{k_m}} \right]. \tag{3.22}$$

The matrix $\nabla H_{map}^T R_{map}^{-1} \nabla H_{map}$ in (3.15) will be of the form

$$\nabla H_{map}^T R_{map}^{-1} \nabla H_{map} = \begin{bmatrix} 0 & 0 \\ 0 & \Omega \end{bmatrix} \tag{3.23}$$

in which all the elements relating to features that are not present in the measurement vector are exactly zero and Ω is a $2m \times 2m$ matrix.

In general, each column (row) of the information matrix will contain at most a constant number of non-zero elements (this constant depends on the density of features and the sensor range), independent of the size of the environment and/or the total number of features. Thus the information matrix contains $O(N)$ non-zero elements, where N is the number of features in the map.

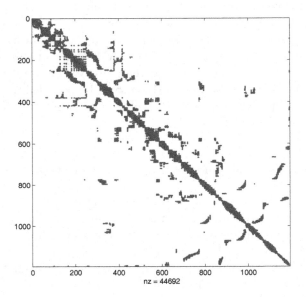

Fig. 3.5   Sparse information matrix obtained by D-SLAM without any reordering of the state vector.

In contrast to many algorithms in the literature, approximation or expanding the state dimension by incorporating intermediate robot poses are not necessary in D-SLAM to achieve the exact sparseness of the information matrix. As will be addressed in Section 3.7, the sparseness of the information matrix leads to significant computational advantages. Figure 3.5 shows a typical information matrix observed during the simulation presented in Section 3.5. The non-zero elements of the information matrix are displayed in black.

3.3   Efficient State and Covariance Recovery

In EIF-based SLAM algorithms, the state vector and the covariance matrix are not computed as a natural part of the estimation process. Instead, the information vector and information matrix are updated in each step.

Although sufficient to describe the estimated state and its uncertainty in the information form, they can not be used directly to perform the observation/process model linearization and data association. Therefore, the recovery of the relevant states and the associated covariances is necessary.

Feature location estimates and the covariances associated with the features in the vicinity of the robot are needed for data association, map update and robot localization. In fact, for mapping, the locations of features that are being observed are required to compute ∇H_{map}, $\nabla H_{map}\hat{X}(k)$ and $H_{map}(\hat{X}(k))$ in (3.15). The locations of previously seen features that are being observed and the associated covariance matrix are required for the localization step. For data association, location estimates and the associated covariances of all the features that are within the sensor range are required. When the number of features is small, these can be simply obtained by (3.11) and computing the inverse of the information matrix. However, when the number of features is large, the computational cost of the inversion will be unacceptable. Therefore, it is crucial to find an efficient method for state and covariance recovery.

The feature location estimate $\hat{X}(k)$ can be recovered by solving the set of sparse linear equations (3.11) rewritten as

$$I(k)\hat{X}(k) = i(k) \tag{3.24}$$

in which $\hat{X}(k)$ is the unknown variable vector and $I(k)$ is the coefficient matrix. It should be noted that this coefficient matrix, $I(k)$, is a symmetric and sparse matrix. Therefore the sparse matrix techniques can be used to make the computation efficient.

Covariance recovery can also be done in a similar manner. Note that

$$I(k)P_i = [0 \cdots 0\ 1\ 0 \cdots 0]^T \tag{3.25}$$

where P_i is the i-th column of the unknown covariance matrix and again the symmetric and sparse information matrix, $I(k)$, is the coefficient matrix. Solving several linear equations of this form, columns of the covariance matrix which correspond to the features that are possibly being observed can be recovered.

It is worth noting that the structure of linear equations (3.24) and (3.25) are similar and they share the same coefficient matrix. This similarity indicates that the same technique for efficient computation can be applied to solving both of these linear equations. Two efficient algorithms for this process are presented in the following. These methods evaluate the recovery of the state estimate and the exact value of the associated covariances. They are therefore different from the method in [Eustice *et al.* (2005c)] used for recovering the conservative bound of the covariance matrix.

3.3.1   *Recovery Using the Preconditioned Conjugated Gradient (PCG) Method*

The Conjugated Gradient (CG) method is a classical iterative method used to solve large sets of linear equations [Barrett *et al.* (1994)]. Each iteration of CG takes $O(N_z)$ operations, where N_z is the number of non-zero entries in the coefficient matrix of the linear equation set. In general, CG requires N iterations to converge, where N is the size of the unknown variable vector.

In the recovery algorithm described in this section, the state estimate and the associated covariances are computed by solving several linear equations using the Preconditioned Conjugated Gradient (PCG) method. The number of iterations required for convergence can be substantially reduced by using PCG. With a good preconditioner, only a few (constant number) iterations are sufficient for the PCG to converge.

A good preconditioner itself requires significant amount of computation in general. However, the coefficient matrix in (3.24) and (3.25) is the symmetric and sparse information matrix. Moreover, the information matrices from two consecutive steps are very similar due to the gradual evolution of the map governed by

$$I(k+1) = I(k) + \nabla H_{map}^T R_{map}^{-1} \nabla H_{map}. \tag{3.26}$$

As discussed in Section 3.2, $\nabla H_{map}^T R_{map}^{-1} \nabla H_{map}$ contains very few non-zero elements. Thus only a few elements in the information matrix are modified in each update step. This special structure of the D-SLAM information matrix leads to an efficient method for the incremental computation of a good preconditioner based on approximate Cholesky Factorization.

3.3.1.1   *Incremental Procedure for Preconditioning*

Let $\nabla H_{map}^T R_{map}^{-1} \nabla H_{map}$ in equation (3.15) be expressed in the form

$$\nabla H_{map}^T R_{map}^{-1} \nabla H_{map} = \begin{bmatrix} 0 & 0 \\ 0 & H_R \end{bmatrix} \tag{3.27}$$

where the upper left element of H_R is non-zero.

The dimension of the square matrix H_R depends on which features are observed at time $k+1$. If all the observed features are present near the bottom of the current state vector, the dimension of H_R is of $O(1)$. On the other hand, if the observed feature set also contains a feature near the top of the state vector, the dimension of H_R is of $O(N)$. The preconditioning process proposed is a function of the dimension of H_R.

Case (i). When the dimension of H_R is less than a threshold n_0[1], the approximate Cholesky Factorization of $I(k)$ is used to construct an approximate Cholesky Factorization of $I(k+1)$ as follows.

Suppose the approximate Cholesky Factorization of $I(k)$ is $\tilde{L}_k$. Let $\tilde{L}_k$ and $I(k)$ be partitioned based on (3.27) as

$$\tilde{L}_k = \begin{bmatrix} \tilde{L}_{11} & 0 \\ \tilde{L}_{21} & \tilde{L}_{22} \end{bmatrix}, \tag{3.28}$$

in which $\tilde{L}_{11}$ and $\tilde{L}_{22}$ are lower triangular matrices, and

$$I(k) = \begin{bmatrix} I_{11} & I_{21}^T \\ I_{21} & I_{22} \end{bmatrix}. \tag{3.29}$$

Let $I(k+1)$ be partitioned as

$$I(k+1) = \begin{bmatrix} I_{11} & I_{21}^T \\ I_{21} & I_{22}^{k+1} \end{bmatrix} = \begin{bmatrix} I_{11} & I_{21}^T \\ I_{21} & I_{22} + H_R \end{bmatrix}. \tag{3.30}$$

As shown in Appendix B.2, an approximate Cholesky Factorization of $I(k+1)$ can be obtained by

$$\tilde{L}_{k+1} = \begin{bmatrix} \tilde{L}_{11} & 0 \\ \tilde{L}_{21} & \tilde{L}_{22}^{k+1} \end{bmatrix} \tag{3.31}$$

where $\tilde{L}_{22}^{k+1}$ is an approximate Cholesky Factorization of the submatrix $H_R + \tilde{L}_{22}\tilde{L}_{22}^T$, which can be computed using the incomplete Cholesky Factorization.

Case (ii). When the dimension of H_R is larger than or equal to n_0, the state vector is reordered based on the distance from each feature to the current robot location or one of the two common features. The feature with the largest distance is placed at the top of the reordered state vector. The

---

[1]Selection of the threshold n_0 is problem dependent. A proper value needs to be set to avoid too much computation in the incomplete Cholesky Factorization of $H_R + \tilde{L}_{22}\tilde{L}_{22}^T$ in Case (i) (if n_0 is too large), and too many reordering steps and the incomplete Cholesky Factorization of $I(k+1)$ in Case (ii) (if n_0 is too small). $n_0 = 100$ is used in the simulation presented in Section 3.5 where the dimension of the final map state vector is 1192.

dimension of H_R that corresponds to the new state vector is a function of the sensor range and will be $O(1)$. Once the information vector and the information matrix are reordered accordingly, an approximate Cholesky Factorization of the whole information matrix needs to be computed to produce the appropriate preconditioner. This is because the preconditioner of the last step can not be used due to reordering. The process of computing the approximate Cholesky Factorization can also be implemented as the incomplete Cholesky Factorization.

3.3.1.2   *Initial Guesses*

Iterative linear equation solvers require an initial guess of the solution vector. Once the preconditioner is well chosen, the initial guess does not significantly influence the required number of iterations.

As the features are stationary, the previous estimates provide good initial guesses for the feature locations. During the experiments and computer simulations, it was seen that zeros appear to be adequate as initial guesses for the elements of the covariance matrix. The performance of the PCG using these initial guesses and the preconditioner calculated from the incremental procedure are shown in Figure 3.20 in Section 3.7.4.

3.3.2   *Recovery Using Complete Cholesky Factorization*

Alternatively, state vector and the covariance matrix can be recovered by solving the sparse linear equations (3.24) and (3.25) using the complete Cholesky Factorization. Suppose the complete Cholesky Factorization of $I(k+1)$ is obtained as $L_{k+1}L_{k+1}^T = I(k+1)$, where L_{k+1} is a lower triangular matrix, the sparse linear equations can be solved by back-substitution.

Moreover, the two consecutive sparse information matrices, $I(k)$ and $I(k+1)$, are very similar due to (3.15). This special structure of the information matrix leads to an efficient method for the incremental computation of complete Cholesky Factorization.

3.3.2.1   *Incremental Procedure for Complete Cholesky Factorization*

Let the first equation in (3.15) be expressed in the form

$$I(k+1) = I(k) + \begin{bmatrix} 0 & 0 \\ 0 & H_R \end{bmatrix}. \tag{3.32}$$

where the upper left element of H_R is non-zero.

The dimension of the square matrix H_R depends on where the features in the current observation are located in the state vector. If all of the features are present near the bottom of the state vector, the dimension of H_R is of $O(1)$. On the other hand, if one of these features is near the top of the state vector, the dimension of H_R is close to the dimension of $I(k+1)$. The Cholesky Factorization process proposed is a function of the dimension of H_R.

Case (i). When the dimension of H_R is less than a threshold n_0[2], the Cholesky Factorization of $I(k)$ is used to construct the Cholesky Factorization of $I(k+1)$ as follows.

Suppose the Cholesky Factorization of $I(k)$ is L_k (a lower triangular matrix). Let L_k and $I(k)$ be partitioned according to (3.32) as

$$L_k = \begin{bmatrix} L_{11} & 0 \\ L_{21} & L_{22} \end{bmatrix}, \quad I(k) = \begin{bmatrix} I_{11} & I_{21}^T \\ I_{21} & I_{22} \end{bmatrix}. \tag{3.33}$$

Using (3.32) and (3.33), $I(k+1)$ can be expressed by

$$I(k+1) = \begin{bmatrix} I_{11} & I_{21}^T \\ I_{21} & I_{22}^{k+1} \end{bmatrix} = \begin{bmatrix} I_{11} & I_{21}^T \\ I_{21} & I_{22} + H_R \end{bmatrix}. \tag{3.34}$$

As shown in Appendix B.1, the Cholesky Factorization of $I(k+1)$ is given by

$$L_{k+1} = \begin{bmatrix} L_{11} & 0 \\ L_{21} & L_{22}^{k+1} \end{bmatrix} \tag{3.35}$$

where L_{22}^{k+1} is the Cholesky Factorization of a low dimensional matrix $H_R + L_{22}L_{22}^T = I_{22}^{k+1} - L_{21}L_{21}^T$.

Case (ii). When the dimension of H_R is larger than or equal to n_0, the state vector is reordered based on the distance from each feature to the current robot location or one of the two common features. The feature with the largest distance is placed at the top of the reordered state vector. The dimension of H_R that corresponds to the new state vector is now very small and will continue to be small until the robot travels a significant distance. Once $i(k+1)$ and $I(k+1)$ are reordered accordingly, a complete Cholesky Factorization of $I(k+1)$ is performed.

---

[2]Similar to the PCG method, selection of the threshold n_0 in this method is also problem dependent. A proper value needs to be set to avoid too much computation in the Cholesky Factorization of $H_R + L_{22}L_{22}^T = I_{22}^{k+1} - L_{21}L_{21}^T$ in Case (i) (if n_0 is too large), and too many reordering steps and Cholesky Factorization of $I(k+1)$ in Case (ii) (if n_0 is too small). $n_0 = 500$ is used in the simulation presented in Section 5.5 where the dimension of the final global map state vector is 2952.

3.3.2.2   *Solution to Sparse Linear Equations Using Complete Cholesky Factorization*

The linear equations (3.24) and (3.25) can be generalized as $I(k+1)X = b$. Once the Cholesky Factorization L_{k+1} is obtained as $L_{k+1}L_{k+1}^T = I(k+1)$, the sparse linear equations $I(k+1)X = b$ can be solved easily by back-substitution to obtain Y in

$$L_{k+1}Y = b \tag{3.36}$$

and then using Y to obtain X in

$$L_{k+1}^T X = Y. \tag{3.37}$$

3.4   Implementation Issues

When the measurements transformation of D-SLAM is introduced in Section 3.1.1, it is assumed that all observations contain measurements to at least two previously seen features. Although this assumption facilitates the presentation of the algorithm, the condition may not be always satisfied, for example when the robot is traveling through a relatively featureless terrain. A method to construct admissible measurements under such scenarios is necessary for D-SLAM to be effective. In practical applications, a reliable and robust data association method is also necessary.

The construction of admissible measurements and a data association algorithm are present in this section.

3.4.1   *Admissible Measurements*

The D-SLAM algorithm described in the previous sections requires that the sensor observes multiple features, including at least two previously seen features at each time step. This condition, while common with a high frequency sensor such as the laser range finder, may not be true when the feature density is low. In such situations, it is possible to combine a sequence of observations to construct an admissible measurement.

Figure 3.6 shows an example where admissible measurements need to be constructed. In the figure, the solid line indicates that the feature is observed from the robot pose; the dashed line indicates that the measurement to the feature is transformed from another robot pose. In this scenario, the robot observes two previously seen features f_1, f_2 and two new features f_3, f_4 at point P_1, observes only one new feature f_5 at point P_2, and only

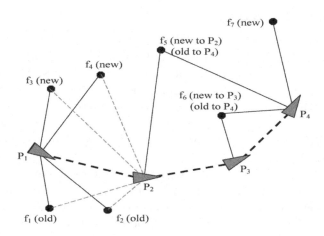

Fig. 3.6 Construction of admissible measurements.

one new feature f_6 at point P_3. Later on at point P_4, it observes features f_5, f_6, f_7. Thus the measurements at P_2 and P_3 are not admissible.

It is, however, possible to combine the measurements made from different points to generate new admissible measurements as follows. Once it is detected that the measurement at point P_2 is not admissible, the update to the map using the observation information from P_1 is removed. This is done by going back to the state estimate before fusing the observation made from P_1. Then a virtual observation from P_2 to f_1, f_2, f_3, f_4 is constructed using the observation from P_1 to f_1, f_2, f_3, f_4 and an estimate of the relative motion of the robot from P_1 to P_2. This is achieved by implementing a small scale EKF SLAM process starting from P_1. The measurements taken from P_1 are fused, and then the robot motion from P_1 to P_2. After that the estimate of this small scale EKF SLAM is transformed to the form of ranges and bearings to construct the virtual observation from P_2. The uncertainty associated with this composite observation is computed using the estimate uncertainty and the transformation equation. The mapping process can now proceed as if features f_1, f_2, f_3, f_4, f_5 are observed from P_2 and no observation is made at P_1. The process is repeated whenever an inadmissible observation is encountered, for example at P_3. This strategy allows D-SLAM to function where features are sparse or one cluster of features are separated from another cluster of features by a region of "featureless" terrain.

3.4.2   Data Association

In the context of SLAM, data association refers to the process of associating the observations with the features that already exist in the map. Many data association algorithms have been proposed for the use of SLAM. Generally speaking, batch data association algorithms (e.g. [Bailey (2002); Neira and Tardos (2001)]) are more robust than the standard maximum likelihood approach [Dissanayake *et al.* (2001)], but have a higher computational cost.

In D-SLAM, data association is required at two instances. One is for the local SLAM (Section 3.1.4.3); the other is for the mapping process of D-SLAM. For the local SLAM process, the standard maximum likelihood approach is adequate. During the D-SLAM mapping process, the correlations between the estimates of the robot pose and feature locations are not maintained. Therefore, maximum likelihood hypotheses of possible associations, assuming that the correlations between the estimates of the robot pose and feature locations are zero, are first generated. These hypotheses are then evaluated using a chi-square test that compares the relative distances and angles among features computed using the measurement vector and the map state vector.

For example, at time step k the robot observes three features f_1, f_2 and f_3. The measurement vector is

$$z_k = \left[r_1, \theta_1, r_2, \theta_2, r_3, \theta_3 \right]^T, \tag{3.38}$$

which contains noise, assumed to be white Gaussian with zero mean and a covariance matrix

$$R_k = diag \left[\sigma_{r_1}^2, \sigma_{\theta_1}^2, \sigma_{r_2}^2, \sigma_{\theta_2}^2 \sigma_{r_3}^2, \sigma_{\theta_3}^2 \right]. \tag{3.39}$$

Suppose hypotheses have been generated that the observations to f_1, f_2, f_3 correspond to three features in the map represented by the states in $X_B = [X_{k1}, X_{k3}, X_{k3}]^T$ respectively. As described in the following, these hypotheses can be verified by using the chi-square test.

Let the relative distances and angles among the features be described as $z_B = [d_{12}, \alpha_{312}, d_{13}]^T$. The physical meaning of the elements of z_B is illustrated in Figure 3.2. These relative quantities can be expressed as a function of the estimate of X_B as

$$z_B = G(\hat{X}_B) = \begin{pmatrix} \sqrt{(\hat{x}_2 - \hat{x}_1)^2 + (\hat{y}_2 - \hat{y}_1)^2} \\ atan2 \left(\frac{\hat{y}_3 - \hat{y}_1}{\hat{x}_3 - \hat{x}_1} \right) - atan2 \left(\frac{\hat{y}_2 - \hat{y}_1}{\hat{x}_2 - \hat{x}_1} \right) \\ \sqrt{(\hat{x}_3 - \hat{x}_1)^2 + (\hat{y}_3 - \hat{y}_1)^2} \end{pmatrix}. \tag{3.40}$$

The associated covariance matrix, P_B, can be computed from the covariances of $\hat{X}_B$ and the function G.

Elements of z_B can also be expressed as a function of the measurement vector as

$$\hat{z}_B = H(z_k) = \begin{bmatrix} \sqrt{(\tilde{x}_2 - \tilde{x}_1)^2 + (\tilde{y}_2 - \tilde{y}_1)^2} \\ atan2\left(\frac{\tilde{y}_3 - \tilde{y}_1}{\tilde{x}_3 - \tilde{x}_1}\right) - atan2\left(\frac{\tilde{y}_2 - \tilde{y}_1}{\tilde{x}_2 - \tilde{x}_1}\right) \\ \sqrt{(\tilde{x}_3 - \tilde{x}_1)^2 + (\tilde{y}_3 - \tilde{y}_1)^2} \end{bmatrix} \tag{3.41}$$

where

$$\tilde{x}_i = r_i \cos\theta_i, \quad \tilde{y}_i = r_i \sin\theta_i, \quad i = 1, 2, 3. \tag{3.42}$$

The associated covariance matrix, R_B, can be computed from the covariances of the measurements and the function H.

Then the innovation and the innovation covariance can be computed by

$$\nu = z_B - \hat{z}_B \tag{3.43}$$

$$S = \nabla G P_B \nabla G^T + \nabla H R_B \nabla H^T \tag{3.44}$$

where ∇G is the Jacobian of function G with respect to X_B evaluated on $\hat{X}_B$ and ∇H is the Jacobian of function H with respect to the elements in the measurements evaluated on the current value of z_k.

The hypotheses that f_1, f_2 and f_3 correspond to the map states in X_B can be accepted if

$$D^2 = \nu^T S^{-1} \nu < \chi^2_{d,\alpha} \tag{3.45}$$

where d is the dimension of ν and α is the desired confidence level.

While the above process does not require the correlation between the estimates of the robot pose and the feature locations, it still needs the location estimates and the associated covariance matrix of the set of features that are potentially being observed. The estimate of this feature set can be extracted from the current map estimate using the current robot location estimate and the sensor range. The presence of special structure in the map, for example the uniform feature spacing, is unlikely to cause incorrect data association in practice because the hypotheses are generated using the current robot pose estimate. The two common features can also be used to validate the data association.

3.5   Computer Simulations

A 2D simulation experiment in MATLAB (Mathwork inc., U.S.A.) using a large number of features was conducted to evaluate D-SLAM and demonstrate its properties. The environment used is a 100 meter square with 1225 features arranged in uniformly spaced rows and columns as shown in Figure 3.7(a).

A robot equipped with a range-bearing sensor with a field of view of 180 degrees and a range of 6 meters and an odometer was used in the simulation. The noise levels of the odometer and sensor measurements were selected to be similar to those of Pioneer DX robots and the SICK laser range finder. In the simulation, the robot starts from the bottom left corner of the square and follows a random trajectory, revisiting many features and closing many loops as seen in Figure 3.7(a).

Figures 3.7(a) and 3.7(b) show the robot trajectory and feature location estimates from traditional EKF SLAM and D-SLAM respectively. The crosses indicate true feature locations and the ellipses indicate 2σ error bounds for feature location estimates. The difference between the feature location estimates from the two algorithms is not distinguishable in the figures. Figures 3.8(a) and 3.8(b) show robot location estimates from traditional EKF SLAM and D-SLAM respectively. The D-SLAM result is consistent in the sense that most of the estimate errors fall within the 2σ error bounds. It can also be noticed that the D-SLAM result is conservative.

Figure 3.9 shows the estimate error and the associated 95% confidence levels for feature 7. It is clear that the estimates are consistent. Figure 3.10 demonstrates the standard deviation of the location estimates for a set of randomly selected features. It shows that the uncertainty of the feature location estimates decreases monotonically. At around 600 seconds, all feature location estimates are improved as a loop is closed.

Figure 3.11 shows the links among the features in the final information matrix. The points in this figure indicate the true feature locations in the environment. Each line linking two features in the environment indicate that there are links in the information matrix about the states corresponding to the locations of these two features. The figure clearly shows that only nearby features can have links in the information matrix.

Figure 3.12(a) shows all the non-zero elements of the information matrix in black. This information matrix is obtained by purposely not using the incremental preconditioning technique in Section 3.3.1.1 to show the sparseness and the original distribution of zero and non-zero elements of

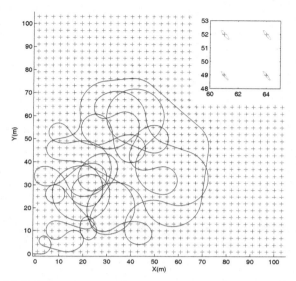

(a) Robot trajectory and the map obtained by traditional EKF SLAM.

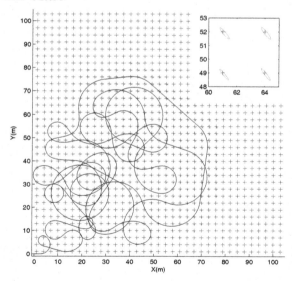

(b) Robot trajectory and the map obtained by D-SLAM

Fig. 3.7   Robot trajectory and the map obtained by traditional EKF SLAM and D-SLAM.

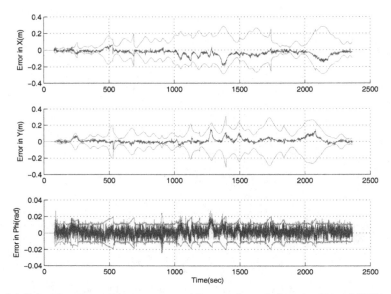

(a) Robot location estimate errors and 2σ error bounds from traditional EKF SLAM

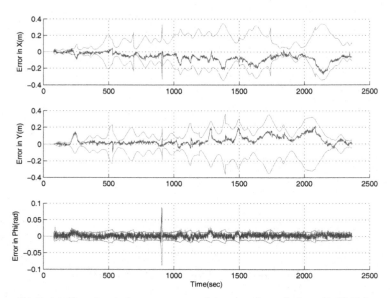

(b) Robot location estimate errors and 2σ error bounds from D-SLAM

Fig. 3.8   Robot location estimate from traditional EKF SLAM and D-SLAM.

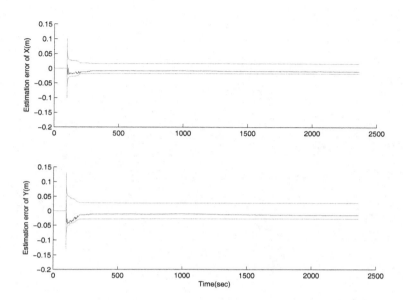

Fig. 3.9   Estimation error of feature 7 from D-SLAM.

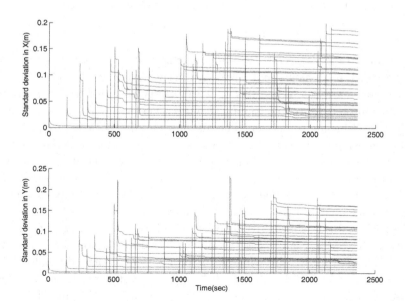

Fig. 3.10   Standard deviation of location estimates of several randomly selected features from D-SLAM.

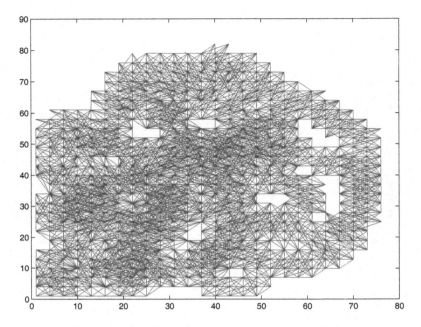

Fig. 3.11   Links in the information matrix from D-SLAM.

the information matrix. It is clear that this matrix is sparse as there are 44692 non-zero elements and 1376172 exactly zero elements. The block diagonal areas are due to features in close vicinity observed together and the off-diagonal terms are due to loop closures where a previously seen feature is reobserved some time later. The information matrix in Figure 3.12(b) is obtained using the incremental preconditioning technique in Section 3.3.1.1. The information matrix is banded due to the reordering of the state vector that occurs from time to time when the incremental preconditioning technique is used. This matrix demonstrates the fact that only the nearby features are linked in the information matrix.

Figure 3.13(a) shows the Cholesky Factorization of the information matrix shown in Figure 3.12(a) which has 128043 non-zero elements. Figure 3.13(b) shows the Cholesky Factorization of the information matrix shown in Figure 3.12(b) which has 168221 non-zero elements. These figures demonstrate that the reordering of the information matrix significantly reduces the number of fill-ins during the Cholesky Factorization.

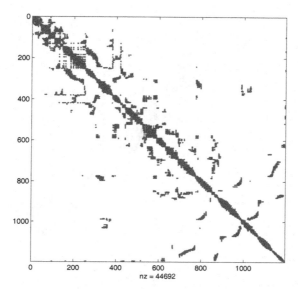

(a) Sparse information matrix obtained by D-SLAM without any reordering of the state vector.

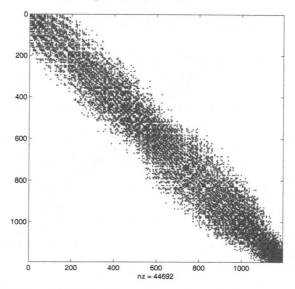

(b) Banded information matrix obtained by D-SLAM with reordering of the state vector.

Fig. 3.12 Sparse information matrix from D-SLAM.

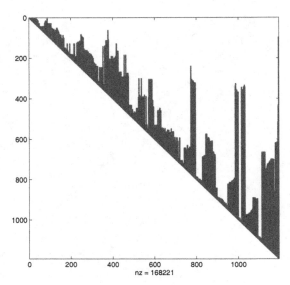

(a) Cholesky Factorization of the information matrix
without any reordering of the state vector.

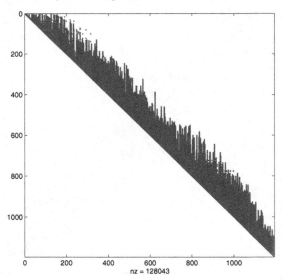

(b) Cholesky Factorization of the information matrix
with reordering of the state vector.

Fig. 3.13   Cholesky Factorization of the information matrix from D-SLAM.

3.6 Experimental Evaluation

3.6.1 *Experiment in a Small Environment*

The Pioneer 2 DX robot was used for the experimental evaluation. This robot is equipped with a laser range finder with a field of view of 180 degrees and an angular resolution of 0.5 degree. The test site was in the laboratory where twelve laser reflector strips were placed in an $8 \times 8m^2$ area. The Player software [Gerkey *et al.* (2003)] was used to collect the control and sensor data from the robot. A MATLAB implementation of D-SLAM was used to process the data and compute the robot pose and feature locations.

Figure 3.14 shows the map obtained from D-SLAM. Figure 3.15 shows the error in the robot location estimate from D-SLAM with respect to that from the traditional EKF SLAM algorithm, which is used as baseline in the comparison. The 2σ error bounds in this figure are computed by combining the variances obtained from the two algorithms. Figures 3.16(a) and 3.16(b) show the 2σ error bounds obtained from D-SLAM and traditional EKF SLAM for the estimates of the robot pose and the location of feature 9 respectively.

Figure 3.15 shows that the localization error in D-SLAM falls within the 2σ error bounds, and this indicates that the estimate is consistent. The map shown in Figure 3.14 is almost as good as that from traditional EKF SLAM in this small area, as can be seen more clearly in Figure 3.16(b). In this figure, the 2σ error bounds from D-SLAM is very close to those from traditional EKF SLAM.

It can be seen from Figure 3.16(a) that the localization result using CI is conservative compared with that from traditional EKF SLAM, as expected. The differences between the 2σ error bounds from D-SLAM and those from traditional EKF SLAM in robot location and orientation estimates are less than 0.05 m and 0.02 rad respectively, which are small compared with the size of the environment and are acceptable.

3.6.2 *Experiment Using the Victoria Park Dataset*

This large scale outdoor dataset, available from Australian Centre of Field Robotics (ACFR) [Guivant and Nebot (2001)], is collected by a standard utility vehicle shown in Figure 3.17(b) which is fitted with dead reckoning sensors and a laser range finder. Figure 3.17(a) presents a satellite image of the park. Information from the laser is processed to extract location of the

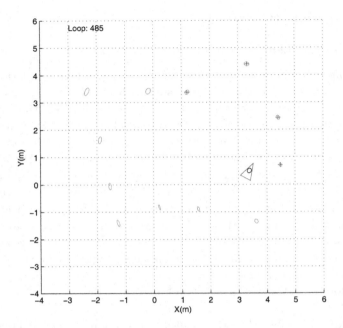

Fig. 3.14   Map obtained by D-SLAM (ellipses indicate 2σ error bounds for feature and robot location estimates; crosses indicate features currently being observed).

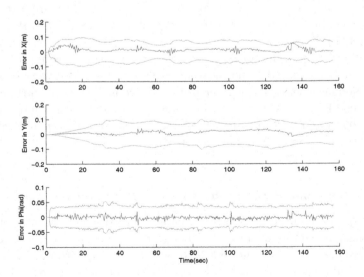

Fig. 3.15   Error of the robot location estimate and associated 2σ error bounds.

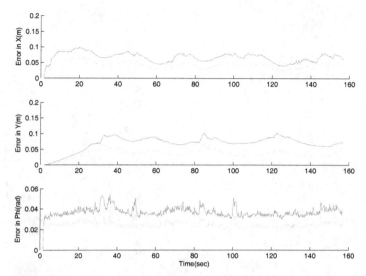

(a) 2σ error bounds of robot location estimate (solid line is from D-SLAM; dashed line is from traditional EKF SLAM).

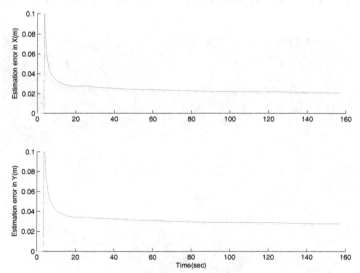

(b) 2σ error bounds for the location estimate of feature 9 (solid line is from D-SLAM; dashed line is from traditional EKF SLAM).

Fig. 3.16 Estimate of robot location and feature 9 from traditional EKF SLAM and D-SLAM.

(a) Victoria Park.

(b) The vehicle.

Fig. 3.17   Victoria park and the vehicle.

trees in the park. Ground truth for this dataset is not available. Therefore it is only possible to comment on the outputs of the two algorithms used, traditional EKF SLAM and D-SLAM. It is important to note that during this practical example, D-SLAM and traditional EKF SLAM behave slightly differently as feature validation and data association strategies are specific to the algorithm being implemented. Therefore some features present in the traditional EKF SLAM map are not present in the D-SLAM map and vice versa.

The estimates of the feature positions and the vehicle trajectories obtained by the D-SLAM algorithm and the traditional EKF SLAM algorithm are shown in Figure 3.18(a). The maximal difference between the two trajectories is around $3m$.

Figure 3.18(b) shows the sparse information matrix of the D-SLAM result. In this scenario, the sensor range is significant with respect to the size of the environment. The information matrix is, therefore, not very sparse. In particular, the high density of non-zero elements in the top-left part of the matrix is caused by many loop closing events near the starting point of the vehicle.

Figure 3.19(a) shows the standard deviations of the vehicle position estimates obtained from D-SLAM and traditional EKF SLAM. It is seen, as expected, that the estimates of the errors from D-SLAM are always higher than those from traditional EKF SLAM but the differences are clearly not very significant.

Figure 3.19(b) compares the uncertainties of the location estimates for two features. These features correspond to the two extremes in terms of difference in final feature location estimate uncertainty between the two algorithms: one corresponds to the smallest difference, and the other corresponds to the largest. Again the estimates of the errors from D-SLAM are always above those from traditional EKF SLAM, confirming that D-SLAM is conservative. However, even in the worst case scenario, the difference in the estimate uncertainty is less than $1m$.

3.7 Computational Complexity

Let the number of features in the map be N and the robot operate on a two-dimensional plane.

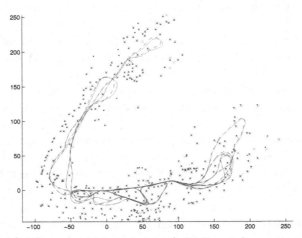

(a) Map and vehicle trajectory (solid line and cross are trajectory and feature location from D-SLAM; dashed line and dot are trajectory and feature location from traditional EKF SLAM).

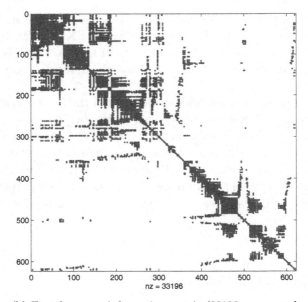

(b) Exactly sparse information matrix (33196 non-zero elements and 353688 exactly zero elements).

Fig. 3.18   Map and exactly sparse information matrix.

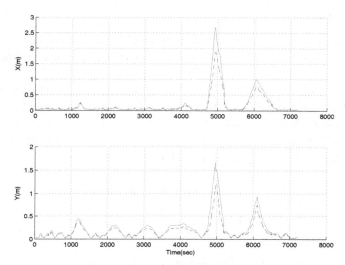

(a) 1σ error bounds for robot position estimate (solid line is from D-SLAM; dashed line is from traditional EKF SLAM).

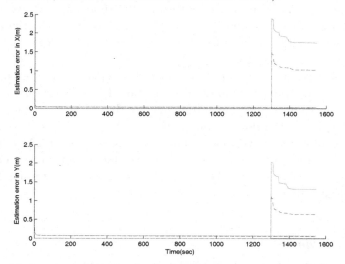

(b) 2σ error bounds for the feature location estimate of the two extreme cases in terms of difference in resulting estimation uncertainty (solid line is from D-SLAM; dashed line is from traditional EKF SLAM).

Fig. 3.19 Estimates of feature locations and robot pose from traditional EKF SLAM and D-SLAM.

3.7.1   *Storage*

Storage is required for the information vector, the recovered state vector, the sparse information matrix and a limited number of columns of the covariance matrix.

In a two-dimensional scenario, the dimension of both the information vector and the recovered state vector is $2N$. Although the information matrix has $O(N^2)$ elements, only $O(N)$ elements are non-zero, thus the cost of storage is $O(N)$. A number of columns of the covariance matrix corresponding to the features in the vicinity of the robot require $O(N)$ storage.

The overall storage requirement is therefore $O(N)$.

3.7.2   *Localization*

In Section 3.1.4.1, the content and dimension of the state vector used in D-SLAM localization are analyzed with respect to the localization flow chart in Figure 3.4. The analysis shows that at any time, the state vector only contains the robot pose and locations of features observed by the robot at the moment. Thus the dimension of the state vector is bounded and can be considered as a constant.

At each localization step, one robot pose estimate is obtained from the local SLAM process through the standard EKF operations; the other robot pose estimate is obtained from solving the "kidnapped robot" problem through a relocation operation using the current observation. The dimension of the state vectors for both these processes are $2M + 3$, where M is the number of features observed at the time instant under consideration. For a large scale map, $M \ll N$. Therefore, the computational cost of localization is $O(1)$.

3.7.3   *Mapping*

Mapping in D-SLAM is formulated in the information form as shown in equation (3.15), which is

$$\begin{aligned}
I(k+1) &= I(k) + \nabla H_{map}^T R_{map}^{-1} \nabla H_{map} \\
i(k+1) &= i(k) + \nabla H_{map}^T R_{map}^{-1} [z_{map}(k+1) \\
&\quad - H_{map}(\hat{X}(k)) + \nabla H_{map}\hat{X}(k)]
\end{aligned} \qquad (3.46)$$

where ∇H_{map} is the Jacobian of the function H_{map} with respect to all the states evaluated on the current state estimate $\hat{X}(k)$.

As discussed in Section 3.2, the Jacobian ∇H_{map} in equation (3.46) is sparse with only some diagonal and off-diagonal elements which are directly related to the currently observed features being non-zero. Therefore ∇H_{map} is sparse with a constant number of non-zero elements. Thus all the operations required for the map update as defined in equation (3.46) require constant computation time when the recovered state vector is available. The prediction step, the computationally demanding stage of an information filter, does not exist in D-SLAM.

Therefore, mapping in D-SLAM is an $O(1)$ time process.

3.7.4 *State and Covariance Recovery*

The major computational cost of D-SLAM is due to the need for recovering the feature location estimates and a number of columns of the associated covariance matrix by solving a set of linear equations. The feature location estimates are needed to compute the Jacobian of the observation model which is required in the map update. A number of columns of the covariance matrix are required for data association. In Section 3.3, two efficient recovery methods are proposed: Preconditioned Conjugated Gradient and the complete Cholesky Factorization together with back-substitution. In this section, the computational cost of the PCG based strategy is analyzed. The recovery method using complete Cholesky Factorization described in Section 3.3.2 has very similar characteristics and is implemented and analyzed in Chapter 5.

As discussed in Section 3.3, state and covariances recovery is achieved by solving the sparse linear equations (3.24) and (3.25), which can be written in the general form

$$A_{n \times n} x_{n \times 1} = b_{n \times 1} \tag{3.47}$$

where x is the unknown variable vector and A is the coefficient matrix.

When solving a set of linear equations in (3.47) using the Conjugated Gradient (CG) method, the computational cost of each iteration is dominated by a matrix-vector product, $A_{n \times n} d_{n \times 1}$, where d is an arbitrary vector updated in each iteration. This product requires $O(n_z)$ operations, where n_z is the number of non-zero entries in matrix $A_{n \times n}$ (see [Shewchuk (1994)]).

In D-SLAM, the matrix A is the sparse information matrix with each column (row) containing at most a constant number of non-zero elements (see Section 3.2). Therefore, matrix A contains $O(N)$ non-zero elements. Thus, the matrix-vector multiplication that is necessary at each iteration of CG requires $O(N)$ operations, and the computational cost of each iteration in CG is $O(N)$. Generally, CG requires N iterations to converge. However, with a good preconditioner, the preconditioned CG (PCG) only requires a few (constant number) iterations to converge. The incremental procedure for preconditioning proposed in Section 3.3.1.1 can produce such a preconditioner at a very small computational cost. It should be noted that the same preconditioner can be used for both state and covariances recovery.

The computational cost in Case (i) presented in Section 3.3.1.1 is $O(1)$ since the size of the submatrix $H_R + \tilde{L}_{22}\tilde{L}_{22}^T$ is less than a constant $n_0 \times n_0$. Case (ii) involves reordering the state vector, the information vector and the associated information matrix, together with an approximate Cholesky Factorization of the reordered sparse information matrix. Therefore Case (ii) will be more computationally expensive as the incremental procedure described in Case (i) can not be used to compute the preconditioner. However, Case (ii) only occurs occasionally[3] as the state vector is reordered according to the distance from each feature to the current robot location. After one such reordering, Case (i) will apply until the robot travels a large distance and observes a feature located near the top of the state vector. Clearly the computational cost of the recovery process is governed by the frequency at which Case (ii) occurs during the mapping process. This is a function of the environment structure. Therefore, a theoretical prediction of the overall computational cost can not be made.

In the simulation presented in Section 3.5, the approximate Cholesky Factorization of the submatrix $H_R + \tilde{L}_{22}\tilde{L}_{22}^T$ in Case (i) and that of the reordered information matrix in Case (ii) described in Section 3.3.1.1 was implemented as the incomplete Cholesky Factorization using MATLAB build-in function "cholinc" with the drop tolerance 10^{-7}. Figure 3.20 shows the time[4] required to recover the state vector and one column of the covariance matrix using PCG together with the time required to compute the precon-

---

[3]The number of times Case (ii) can occur depends on the parameter n_0, the sensor range, the density of features and the robot trajectory. If there is no loop closure, Case (ii) will never happen. In the simulation presented in Section 3.5 which contains many loop closures, Case (ii) occurs in 73 out of total 7876 steps.

[4]All time measurements in this chapter are performed on a desktop with a 2.8GHz Intel Pentium processor, 1GB of RAM and running Windows. All programs are written in MATLAB.

ditioner using the incremental procedure described in Section 3.3.1.1 as a function of the number of features in the state. In Figure 3.20 the average of the computational time required for all these steps is used for the cases in which the number of features is identical during multiple steps. The ratio between the computational cost and the number of features is seen to be a constant when the number of features is large. This indicates that although a theoretical prediction for the computational cost can not be made, the computational cost of PCG (recovering the state and/or one column of the covariance matrix) and preconditioning are both $O(N)$ for large maps. Figure 3.21 shows the average number of PCG iterations as a function of the number of features, where it is seen that PCG takes at most 2 iterations to converge. This may not be true in some special circumstances.

Figure 3.22 compares the average time required for the Cholesky Factorization using direct method used in [Dellaert (2005)] for incremental $\sqrt{SAM}$ and the average time for the Cholesky Factorization using the incremental method described in Section 3.3.1.1. It is clear that the computational saving achieved by the incremental method is significant.

During the execution of the approximate Cholesky Factorization, there may be entries which change from a zero to a non-zero value, which are

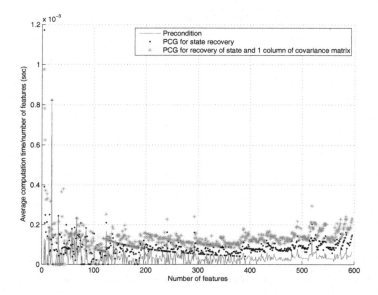

Fig. 3.20 Average time required for preconditioning and PCG divided by the number of features.

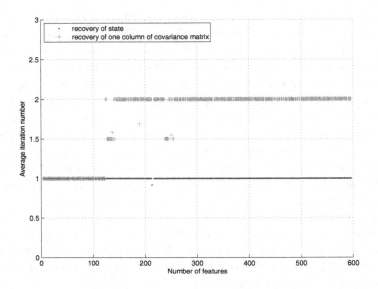

Fig. 3.21   Average number of iterations required for PCG convergence.

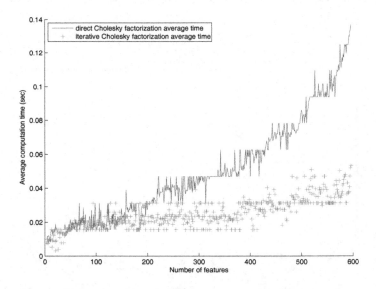

Fig. 3.22   Comparison of the average time required for the Cholesky Factorization.

referred to as "fill-in". Fill-ins reduce the sparseness of the coefficient matrix. It is well known that the extent of fill-ins is highly related to the order in which the variables are eliminated. The reordering of the state vector according to distance as described in Case (ii) changes the order of this variable elimination process. Simulation results (Figures 3.13(a) and 3.13(b)) show that the reordering process described in Case (ii) significantly reduces the number of fill-ins. This reordering has a similar effect to that shown in [Dellaert (2005)].

3.8 Consistency of D-SLAM

It is well known that linearization errors can lead to inconsistent estimates in EKF/EIF SLAM. With the process model and measurement model being nonlinear, the EKF/EIF SLAM is a nonlinear estimation problem. The Jacobians of the process and observation models need to be computed at each step using the current estimate values, which causes the linearization errors. Julier and Uhlmann (2001b) showed theoretically that when the robot is stationary with non-zero initial pose uncertainty and observes a new feature repeatedly, the EKF SLAM algorithm always produces inconsistent robot location and map estimates. It was also shown through simulations, that the same inconsistency problem appears when the robot is moving.

As an EIF-based SLAM algorithm, D-SLAM is not exempt from the inconsistencies due to linearization errors. Linearization is needed for the observation model in the map update step as defined by (3.15), the nonlinear function used to solve the "kidnapped robot" problem for localization in (3.17), as well as the robot process model and the observation model in the local SLAM process for localization described in Section 3.1.4.3.

Another source of inconsistency in EKF/EIF SLAM is the manipulation of the estimation process, for example, the approximate sparsification used in SEIF [Thrun *et al.* (2004a)]. D-SLAM does not require such an approximation. Using CI to combine the two robot location estimates avoids possible inconsistency due to information reuse during localization. Although the robot location computed by the process described in Section 3.1.4.4 is correlated to the map, this correlation does not affect the mapping process as the information about the robot location is not exploited during mapping. Thus the D-SLAM algorithm does not contain any approximations beyond the linearization that can lead to estimator inconsistency.

3.9   Bibliographical Remarks

An empirical finding that the normalized information matrix obtained when the SLAM problem is formulated in the information form is approximately sparse motivates the Sparse Extended Information Filter (SEIF) by [Thrun *et al.* (2004a)]. Sparsification in SEIF essentially removes the weak links in the information matrix by setting elements that are smaller than a given threshold to zero, while strengthening other links to make up for the effect of this change. This results in an information matrix with many zero elements and consequently the prediction and update steps can be performed in constant time. However, Eustice *et al.* (2005b) demonstrated that the sparsification process used in SEIF is equivalent to assuming conditional independence between the robot pose and a subset of the features. As this assumption is incorrect in general, SEIF is stochastically inconsistent.

When all the robot poses from which the measurements are made and all the features are included in the state vector, the SLAM problem becomes a static estimation problem. This situation is discussed in [Frese (2005)]. The observation information only links the observed features and the robot pose at which the observation is made, while the robot process model only links consecutive robot poses. Thus most off-diagonal elements in the information matrix are exactly zero and the information matrix is exactly sparse. A state vector that includes all robot poses is also used in Square Root SAM [Dellaert and Kaess (2006)]. In Square Root SAM, the SLAM problem is formulated as a linearized least squares problem and is solved by factorizing the smoothing information matrix using Cholesky or QR Factorization in a batch or incremental manner. In each step the estimates of all previous robot poses are updated together with the feature location estimates through the smoothing process. The updated robot pose estimates are used in the following steps to recompute Jacobians for the measurement equations of old measurements. Thus the linearization error is reduced. Therefore, the SAM algorithm is less prone to estimator inconsistency that can arise due to linearization errors than all other EIF algorithms discussed in this book. However, the sparseness of the information matrix is achieved through increasing the state dimension, which keeps increasing even when the robot is revisiting previously explored regions. Therefore, the computational cost increases over time and is not bounded by the total number of features in the environment.

By computing the relationship between two consecutive robot poses using the observations made at these poses, the SLAM problem can be formulated using a state vector that contains only the robot poses. In this scenario, the resulting information matrix is exactly sparse as shown in the Exactly Sparse Delayed State Filter (ESDSF) [Eustice *et al.* (2006)]. The key advantage of ESDSF is that it is suitable for the scenarios where features are difficult to extract or the number of features is too large compared to the number of robot poses. Effectiveness of ESDSF is demonstrated by the excellent maps shown in [Eustice *et al.* (2006)]. However, the resulting "map" is only an alignment of a sequence of observations (such as images or laser scans). There are no statistical map updates, thus improvements to the state estimates achieved through feature location updates in traditional EKF SLAM are not present. The extent of the information loss due to this has not yet been analyzed.

The Exactly Sparse Extended Information Filter (ESEIF) by [Walter *et al.* (2007)] achieves a sparse information matrix by periodically "kidnapping" and "relocating" the robot. ESEIF exploits the fact that when the robot pose is marginalized out from the state vector, new links will only be built up among the features that are previously linked with the robot pose in the information matrix. The set of features that are linked with the robot pose are called "active features". Thus the information matrix will be sparse if the number of "active features" is bounded. In contrast to the sparsification process in SEIF, ESEIF controls the number of "active features" by "kidnapping" the robot when the number of "active features" is about to become larger than a predefined threshold Γ_a. This is followed by "relocating" the robot using a set of selected measurements. Thus, the information matrix is kept sparse without any approximation that can lead to inconsistency. There is some information loss in ESEIF due to "kidnapping" and "relocating" the robot. The extent of the information loss is mainly related to the frequency that the threshold Γ_a is exceeded as well as the ratio between the measurement and process noises.

In recent years, a number of researchers have also discussed the possibility of decoupling the mapping and localization processes in SLAM in order to gain computational efficiency. Since the observations made from the robot contain the relative locations between the robot and features, a natural way to decouple mapping and localization is to extract the information about the relative locations among the features and then construct a relative map.

Csorba *et al.* (1997) introduced the Relative Filter (REL filter), in which the state vector to be estimated contains relative distances and angles between features. Due to the absence of robot location in the state vector, the errors of different relative states are uncorrelated, which makes the update process only affect the relative states that influence the current observation. Thus the algorithm has $O(N)$ storage requirement and $O(1)$ computational cost, where N is the number of elements in the state. However, due to the presence of redundant elements in the state vector, the estimated relative map can not guarantee to form a geometrically consistent absolute map.

Newman (2000) enforced geometric constraints of the map in REL filter by projecting the relative state estimate to a lower dimension space and thus developed the Geometric Projection Filter (GPF), which produces geometrically consistent absolute maps. GPF inherits the computational advantages of REL filter in building the relative map while overcoming its disadvantage. However, the enforcement of constraints requires more computation and needs to be performed properly by, for example, careful constraint formulation, scheduling of constraint application or the use of submaps.

The work from [Deans and Hebert (2000)] bears the same principle as REL filter and GPF, making use of relative distances and angles which are invariant to rotation and translation. In this algorithm, bearing only SLAM is considered. The geometric consistency constraints are enforced by applying Levenberg-Marquardt optimization, but this requires a significant computational effort. Pradalier and Sekhavat (2003) combined the algorithm by [Deans and Hebert (2000)] with the concept of correspondence graphs. In this work, several issues that were not explicitly addressed in previous relative map algorithms are addressed, such as localization and data association.

Martinelli *et al.* (2004) made use of relative maps where the map state contains only relative distances among the features, which are invariants under translation and rotation. The structure of the covariance matrix is kept sparse by maintaining a state vector with redundant elements. As the relationships among the map elements are not enforced, for large scale problems the map becomes complex and difficult to use. However, if the constraints that enforce these relationships are applied, the simple structure of the covariance matrix is destroyed, leading to an increased computational complexity.

All above decoupling algorithms use the EKF to estimate a relative map state with redundant elements. In contrast, D-SLAM presented in this chapter achieves decoupling of localization and mapping using the EIF and an absolute map state with no redundancy. The information matrix is exactly sparse even in the presence of loop closure events. This leads to great computational advantages when an EIF is used in the estimation process.

3.10   Summary

In this chapter, a decoupled SLAM algorithm, D-SLAM, is described. The localization and mapping are separate processes although they are performed simultaneously in D-SLAM. This new algorithm is based on a method to recast the observation vector such that the information about the map is extracted independent of the robot location.

Although the robot location is not incorporated in the state vector used in mapping, correlations among the feature location estimates are still preserved. Thus the estimates of all the feature locations are improved using information from one local observation.

The significant advantages gained are that there is no prediction step for the mapping, and that the information matrix associated with mapping is exactly sparse and only the features that are observed simultaneously are linked through the information matrix.

Two efficient methods to recover the feature location estimates and the associated covariances without any approximation are presented. One method is based on solving linear equations exploiting the sparse structure of the information matrix by the iterative method PCG. An incremental Cholesky Factorization procedure to produce the preconditioner is developed from the similarity of the information matrices of successive steps, which speeds up the converge rate of CG. The other method is based on the complete Cholesky Factorization. The factorization process is made efficient also by exploiting the similarity of the information matrices of successive steps. With the factorization, the linear equations can be solved by back-substitution.

Total storage and computational cost for each step in D-SLAM are analyzed in terms of the total number of features in the state. The computer simulations demonstrate that the sparse structure of the information matrix and the efficient recovery algorithms presented in Section 3.3 lead to a linear computational cost with respect to the size of the map.

It is shown that D-SLAM does not introduce inconsistencies beyond those resulting from the linearization errors in achieving the exactly sparse information matrix. However, this process causes some information loss. This is due to the fact that not all the information about the state contained in the measurements is exploited during the estimation process.

Chapter 4

D-SLAM Local Map Joining Filter

This chapter presents a local map based SLAM algorithm for the mapping of large scale environments using the D-SLAM framework [Huang *et al.* (2009)]. The local maps can be generated by traditional EKF SLAM. Relative position information among the features in the local map is first extracted. Then the D-SLAM framework is used to fuse the relative information and build a global map.

In the global map building process, the input is the relative location information among features in the local map. The local maps are to be constructed such that the features present in each local map is a small fraction of the features present in the global map. Also the robot pose is not included in the global state vector. Thus, the conditions set in Section 2.4 for the sparseness of the information matrix are satisfied during the global map construction process. Similar to the analysis shown in Section 3.7, the sparse information matrix of the global map leads to significant computational benefits.

This local map based SLAM algorithm eliminates a significant component of the information loss of D-SLAM. This is because at a local level all the information from the process model is fused into the local state estimate. However, in computing the relative information among features in the local map, the robot pose estimate is ignored, still causing some information loss.

This chapter is organized as the follows. In Section 4.1, the structure of the D-SLAM Local Map Joining Filter is stated. Section 4.2 addresses how to obtain relative location information from local maps, and Section 4.3 discusses how to use this information to update the global map estimate. Section 4.4 addresses some implementation issues such as data association, global state and associated covariance matrix recovery, etc. In Section 4.5, computational complexity of the algorithm is analyzed. Simulation results are presented in Section 4.6 and experiment results using the Victoria park dataset are presented in Section 4.7 to evaluate the algorithm. In Section 4.8, the work in the literature on local map based SLAM is reviewed. The chapter is summarized in Section 4.9.

4.1   Structure of D-SLAM Local Map Joining Filter

The D-SLAM Local Map Joining Filter consists of two levels as shown in Figure 4.1 and a global map is maintained explicitly. The aim of this algorithm is to build a global map of the environment, which contains all feature location estimates with respect to a global coordinate frame.

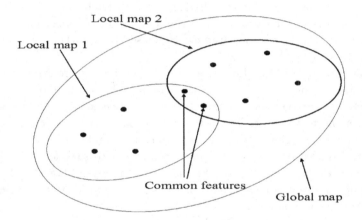

Fig. 4.1   The global map and local maps.

On the local level, the local map is constructed such that it contains at least two features that are present in the existing global map. Traditional EKF SLAM is performed to summarize information from the observations of local features and the robot process in the local map estimate. The resulting local map contains feature location estimates with respect to the local coordinate frame. This process of building a local map is continuous as the robot moves and it is terminated immediately when a decision is made to start another local map. Since the area under consideration during local map building is small, the uncertainty of the map estimate and robot location estimate is comparatively small, thus it is expected that the local map estimates are not significantly influenced by the presence of linearization errors, which can create problems during large scale SLAM [Bailey *et al.* (2006); Huang and Dissanayake (2007)].

On the global level, the D-SLAM framework described in Chapter 3 is used to fuse the local maps into a global map. For each local map, the relative information about feature locations is extracted from the state estimate in the form of feature-feature distances and the angles among three features. These relative quantities are independent of the coordinate frame. They can be regarded as a collection of observations about relative feature locations, and thus can be fused in the same manner as the D-SLAM mapping algorithm. Similar to the Constrained Local Submap Filter [Williams (2001)], this operation is not frequent and only happens when a local map is finished.

4.1.1 *State Vectors*

The state vector for the global map contains the absolute location of all the features with respect to the coordinate frame defined by the initial robot pose, with the coordinate origin at the robot initial location and the x-axis along the initial robot heading. Let the total number of features be N, then the state vector for the global map is, denoted using the superscript 'G',

$$
X^G = \begin{bmatrix} X_1^G \\ X_2^G \\ \vdots \\ X_N^G \end{bmatrix} = \begin{bmatrix} x_1^G \\ y_1^G \\ x_2^G \\ y_2^G \\ \vdots \\ x_N^G \\ y_N^G \end{bmatrix} \tag{4.1}
$$

where $X_i^G = [x_i^G, y_i^G]^T (i = 1 \cdots N)$ defines the coordinate of the i-th feature location.

The global map is built based on the information from a set of local maps. The state vector for each local map contains the robot final pose and the coordinates of all the local feature locations in the local coordinate frame. The local coordinate frame is defined by the robot pose when the local map is started, with the coordinate origin at the robot location and x-axis along the robot heading. Suppose the number of features in a local map is n, then the state vector for the local map is, denoted using the superscript 'L',

$$X^L = \begin{bmatrix} X_r^L \\ X_1^L \\ X_2^L \\ \vdots \\ X_n^L \end{bmatrix} = \begin{bmatrix} x_r^L \\ y_r^L \\ \phi_r^L \\ x_1^L \\ y_1^L \\ x_2^L \\ y_2^L \\ \vdots \\ x_n^L \\ y_n^L \end{bmatrix} \qquad (4.2)$$

where $X_r = [x_r^L, y_r^L, \phi_r^L]^T$ defines the coordinate of the robot pose and $X_i^L = [x_i^L, y_i^L]^T (i = 1 \cdots N)$ defines the coordinate of the i-th feature location.

Note that robot poses are not included in the global map state. While in the local map, the robot pose is present in the state vector as the local map is constructed using the traditional EKF SLAM.

4.1.2   Relative Information Relating Feature Locations

Extracting relative feature location information from the estimate of local maps facilitates the process of fusing the local maps into the global map. These relative quantities about features in the local maps can be easily expressed by the global map states, which are global locations of features. The form of the expression is similar to the observation function used in D-SLAM. Note the feature locations in the local map coordinate frame can not be expressed directly by the global map states, since the robot pose is not maintained in the global state vector.

In order to achieve efficiency and avoid inconsistency, the operation to extract the relative information should meet two criteria: 1) extract as much information as possible; 2) avoid the reuse of information.

4.1.3   Combining Local Maps Using Relative Information

Figure 4.1 illustrates the process of fusing the relative information in local maps into a global map. The large oval indicates the global map and the small ovals illustrate two successive local maps. The dots indicate features in local maps. The dots in the overlap of the two small ovals indicate features common to the two successive local maps.

The key difference between the fusing algorithm proposed in this chapter and the existing local map joining methods (e.g. [Tardos *et al.* (2002); Williams (2001)]) is that in this algorithm "relative location information" is used to update the "absolute map". This allows the use of the D-SLAM framework described in Chapter 3 and thus a significant reduction in computational cost can be achieved.

In order that the relative feature location information extracted from a local map can be fused into the global map, certain amount of overlap between the existing global map and the local map under consideration is necessary. For two-dimensional mapping, the local map should contain at least two previously observed features. These overlapping features are needed to find the correspondence between features in the local and global maps. A strategy for overcoming this limitation will be presented in Chapter 5.

4.2 Obtaining Relative Location Information in Local Maps

This section describes the procedure for extracting relative feature location information from a local map.

4.2.1 *Generating a Local Map*

Traditional EKF SLAM is used to obtain the local map. The final result is the estimate of the state vector containing the final robot pose and the coordinates of local features

$$
\hat{X}^L = \begin{bmatrix} \hat{x}_r^L \\ \hat{y}_r^L \\ \hat{\phi}_r^L \\ \hat{x}_1^L \\ \hat{y}_1^L \\ \vdots \\ \hat{x}_n^L \\ \hat{y}_n^L \end{bmatrix}
\tag{4.3}
$$

and the associated covariance matrix

$$\mathbf{P} = \begin{bmatrix} \mathbf{P}^L_{rr} & \mathbf{P}^L_{rf} \\ (\mathbf{P}^L_{rf})^T & \mathbf{P}^L_{ff} \end{bmatrix}. \tag{4.4}$$

4.2.2   Obtaining Relative Location Information in the Local Map

Before introducing the method to obtain the relative feature location information, it is necessary to discuss what types of relative quantities can meet the criteria set out in Section 4.1.2.

Suppose there are five features $f_1, f_2, \cdots, f_5$ in the local map. The location of feature f_i is (x_i, y_i), $i = 1, \cdots, 5$. Removing the robot pose from the local map state, the state vector of the local map is given by

$$\begin{bmatrix} x_1 \\ y_1 \\ \vdots \\ x_5 \\ y_5 \end{bmatrix}. \tag{4.5}$$

The following two vectors encode the information contained in (4.5)

$$\begin{bmatrix} x_2 - x_1 \\ y_2 - y_1 \\ \vdots \\ x_5 - x_1 \\ y_5 - y_1 \\ x_3 - x_2 \\ y_3 - y_2 \\ \vdots \\ x_5 - x_4 \\ y_5 - y_4 \end{bmatrix}, \begin{bmatrix} d_{12} \\ d_{13} \\ d_{14} \\ d_{15} \\ d_{23} \\ d_{24} \\ d_{25} \\ d_{34} \\ d_{35} \\ d_{45} \end{bmatrix}. \tag{4.6}$$

However, the elements of these two state vectors are not independent. For example, in the first vector in (4.6), $x_2 - x_1$ can be expressed by $x_3 - x_1$ and $x_3 - x_2$. This is clearly not acceptable in building the global map.

Figures 4.2 and 4.3 illustrate two ways in which the relative feature location information can be presented without any redundancy.

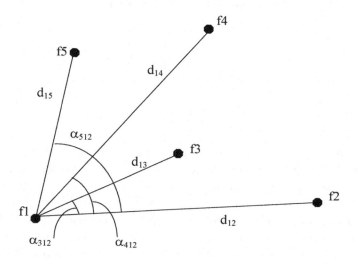

Fig. 4.2   Relative location information version 1.

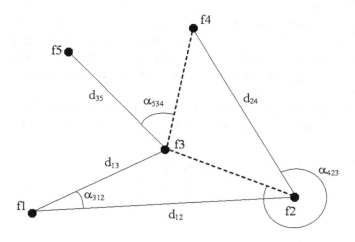

Fig. 4.3   Relative location information version 2.

The state vector corresponding to Figure 4.2 is:

$$\begin{bmatrix} d_{12} \\ \alpha_{312} \\ d_{13} \\ \alpha_{412} \\ d_{14} \\ \alpha_{512} \\ d_{15} \end{bmatrix} . \tag{4.7}$$

In this vector, the relative location information of features is expressed as relative distances between two features and relative angles among three features. For convenience, the feature f_1 is chosen as the base point and the vector $\overrightarrow{f_1 f_2}$ is chosen as the base vector. The relative information relating feature locations includes: 1) distances from all other features to f_1, and 2) angles from all other vectors, $\overrightarrow{f_1 f_i}(i = 3, 4, 5)$, with respect to vector $\overrightarrow{f_1 f_2}$. These relative quantities can be regarded as the polar coordinates in the polar coordinate frame defined by the origin f_1 and the x-axis along vector $\overrightarrow{f_1 f_2}$.

Although the amount of information in the vector of relative quantities in (4.7) is not equal to that in the absolute feature location estimates, the relative distances and angles are independent of the absolute location of features. In other words, they are invariant to shift and rotation of the coordinate frame. This allows fusing the information contained in this vector into the global map without using any coordinate transformation operation, as long as the correspondences between features in the local and global maps are available.

Another form of expressing the relative location information without redundancy is shown in Figure 4.3. The corresponding state vector is:

$$\begin{bmatrix} d_{12} \\ \alpha_{312} \\ d_{13} \\ \alpha_{423} \\ d_{24} \\ \alpha_{534} \\ d_{35} \end{bmatrix} . \tag{4.8}$$

In this vector, the relative location information is independent of the coordinate frame as it is also expressed by relative distances and angles among features. These relative quantities are equivalent to those in (4.7).

Both forms of relative information presented above are suitable for the fusion process. In the algorithm proposed in this chapter, the first form illustrated in Figure 4.2 is used as it is simpler to construct and use.

Suppose the local map estimate $\hat{X}^L$ is available and features f_1 and f_2 are present in the existing global map. The relative distances and angles from other features with respect to features f_1 and f_2 are denoted as

$$z_{map} = \begin{bmatrix} d_{12} \\ \alpha_{312} \\ d_{13} \\ \vdots \\ \alpha_{n12} \\ d_{1n} \end{bmatrix} \tag{4.9}$$

where z_{map} has the same denotation as the transformed measurements in Chapter 3, but the meaning is different. The vector z_{map} can be obtained from the local map estimate $\hat{X}^L$ defined in (4.3) using

$$z_{map} = T_{map}(\hat{X}^L) \tag{4.10}$$

where

$$T_{map}(\hat{X}^L) = \begin{bmatrix} \sqrt{(\hat{x}_2^L - \hat{x}_1^L)^2 + (\hat{y}_2^L - \hat{y}_1^L)^2} \\ atan2\left(\frac{\hat{y}_3^L - \hat{y}_1^L}{\hat{x}_3^L - \hat{x}_1^L}\right) - atan2\left(\frac{\hat{y}_2^L - \hat{y}_1^L}{\hat{x}_2^L - \hat{x}_1^L}\right) \\ \sqrt{(\hat{x}_3^L - \hat{x}_1^L)^2 + (\hat{y}_3^L - \hat{y}_1^L)^2} \\ \vdots \\ atan2\left(\frac{\hat{y}_n^L - \hat{y}_1^L}{\hat{x}_n^L - \hat{x}_1^L}\right) - atan2\left(\frac{\hat{y}_2^L - \hat{y}_1^L}{\hat{x}_2^L - \hat{x}_1^L}\right) \\ \sqrt{(\hat{x}_n^L - \hat{x}_1^L)^2 + (\hat{y}_n^L - \hat{y}_1^L)^2} \end{bmatrix}. \tag{4.11}$$

The corresponding covariance matrix of the uncertainty of these relative quantities can be computed from the covariance matrix of the local map in (4.4) using the relationship between the local map and the relative information described in (4.11) as follows

$$R_{map} = \nabla T_{map} \mathbf{P} \nabla T_{map}^T \tag{4.12}$$

where ∇T_{map} is the Jacobian of the function T_{map} with respect to the local map state evaluated on the estimate $\hat{X}^L$.

The relative information extraction process is similar to the measurement transformation step in the original D-SLAM algorithm described in Section 3.1.1.2. However, in contrast to the original D-SLAM, on the local level, the information contained in the odometry measurements from the robot is exploited during the local map building process. Thus a significant amount of the information loss present in the original D-SLAM is avoided.

4.3   Global Map Update

With the relative feature location information extracted, the local map can be treated as a collection of observations and used to update the existing global map. The fusion process is performed using the D-SLAM framework.

4.3.1   *Measurement Model*

The relative location information can be expressed as a function of the global map state vector (with noise)

$$z_{map} = [d_{12}, \alpha_{312}, d_{13}, \cdots, \alpha_{n12}, d_{1n}]^T = H_{map}(X^G) + w_{map} \qquad (4.13)$$

where

$$H_{map}(X^G) = \begin{pmatrix} \sqrt{(x_2^G - x_1^G)^2 + (y_2^G - y_1^G)^2} \\ atan2\left(\frac{y_3^G - y_1^G}{x_3^G - x_1^G}\right) - atan2\left(\frac{y_2^G - y_1^G}{x_2^G - x_1^G}\right) \\ \sqrt{(x_3^G - x_1^G)^2 + (y_3^G - y_1^G)^2} \\ \vdots \\ atan2\left(\frac{y_n^G - y_1^G}{x_n^G - x_1^G}\right) - atan2\left(\frac{y_2^G - y_1^G}{x_2^G - x_1^G}\right) \\ \sqrt{(x_n^G - x_1^G)^2 + (y_n^G - y_1^G)^2} \end{pmatrix} \qquad (4.14)$$

and w_{map} is the Gaussian measurement noise with zero mean and a covariance matrix R_{map}. The similarity between equations (4.11) and (4.14) is due to the fact that the relative local information, which consists of relative distances and angles among features, is independent of the coordinate frame.

4.3.2   *Updating the Global Map*

The EIF is used to perform the global map update. Let k denote the step when the k-th local map is being fused into the global map. When $k = 1$, the global map is just the same as the first local map without the robot pose.

Let $i(k)$ represent the information vector and $I(k)$ represent the associated information matrix. The estimated state vector and the information vector are related through

$$i(k) = I(k)\hat{X}^G(k). \tag{4.15}$$

After data association, the correspondence among the features in the local map and the global map are identified. Then the new features can be initialized and the global map can be updated.

As shown in Figure 4.4, the new feature f_j can be initialized by the estimates of the two common features in the global map $\hat{X}_1^G = [\hat{x}_1^G, \hat{y}_1^G]^T$ and $\hat{X}_2^G = [\hat{x}_2^G, \hat{y}_2^G]^T$, together with α_{j12}, d_{1j} in z_{map} as

$$\begin{aligned}
\hat{x}_j^G &= \hat{x}_1^G + d_{1j}\cos(atan2(\tfrac{\hat{y}_2^G - \hat{y}_1^G}{\hat{x}_2^G - \hat{x}_1^G}) + \alpha_{j12}) \\
\hat{y}_j^G &= \hat{y}_1^G + d_{1j}\sin(atan2(\tfrac{\hat{y}_2^G - \hat{y}_1^G}{\hat{x}_2^G - \hat{x}_1^G}) + \alpha_{j12}).
\end{aligned} \tag{4.16}$$

For clarity, the new state estimate is still denoted as $\hat{X}^G(k)$. The information vector $i(k)$ and the associated information matrix $I(k)$ are augmented by adding zero elements corresponding to the new features.

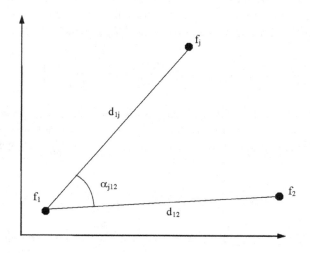

Fig. 4.4   Initialize the new feature in the global map.

Now z_{map} can be used to update the information vector and the information matrix:

$$
\begin{aligned}
I(k+1) &= I(k) + \nabla H_{map}^T R_{map}^{-1} \nabla H_{map} \\
i(k+1) &= i(k) + \nabla H_{map}^T R_{map}^{-1} [z_{map}(k+1) \\
&\quad - H_{map}(\hat{X}^G(k)) + \nabla H_{map} \hat{X}^G(k)]
\end{aligned}
\tag{4.17}
$$

where ∇H_{map} is the Jacobian of the function H_{map} with respect to all the states evaluated on the current state estimate $\hat{X}^G(k)$.

4.3.3   *Sparse Information Matrix*

Since the size of the local map is constructed to be small compared with the global map, only a small fraction of the features in the global map are present in the local map. Thus, the relative feature location information extracted from local map only relates to a limited number of states in the global map state vector. Since the robot poses are not present in the global state vector, the prediction step does not exist. Therefore, the conditions set in Section 2.4 are fulfilled and the information matrix of the global map is sparse.

The sparseness of the global map information matrix can be illustrated by examining the structure of the matrix $\nabla H_{map}^T R_{map}^{-1} \nabla H_{map}$ in (4.17). Since the relative distances and angles only involve a small fraction of the features in the global map, The Jacobian ∇H_{map} will be

$$
\nabla H_{map} = \left[0, \cdots, 0, \frac{\partial H_{map}}{\partial X_{k_1}^G}, \cdots, \frac{\partial H_{map}}{\partial X_{k_n}^G} \right]
\tag{4.18}
$$

where $X_{k_i}, (1 \le i \le n)$ are the global states for the features present in the current local map which are reordered such that these global states are at the bottom of the global state vector. Obviously, $\nabla H_{map}^T R_{map}^{-1} \nabla H_{map}$ will be in the form of

$$
\nabla H_{map}^T R_{map}^{-1} \nabla H_{map} = \begin{bmatrix} 0 & 0 \\ 0 & H_R \end{bmatrix}.
\tag{4.19}
$$

The matrix $\nabla H_{map}^T R_{map}^{-1} \nabla H_{map}$ in (4.17) is sparse with the elements relating to the features that are not present in the local map being exactly zero. As the map update only contains addition operations, the information matrix will be sparse.

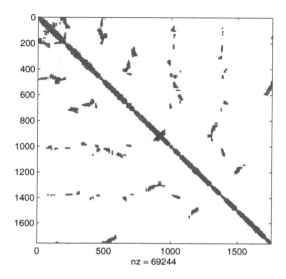

Fig. 4.5   Sparse information matrix without any reordering.

Figure 4.5 shows the sparse information matrix without any reordering in the simulation presented in Section 4.6.2.

4.4   Implementation Issues

This section deals with some implementation issues that need to be addressed so that the proposed algorithm can be used in a practical situation. In particular, estimating the robot location, data association, state and covariance recovery as well as the constraint on the local map that need to be satisfied are addressed in the following.

4.4.1   *Robot Localization*

Knowledge of the robot location in both global and local coordinate frames may be essential for navigating a robot in an unknown environment.

On the local level, the robot location estimate is directly available from the traditional EKF SLAM process within the local coordinate frame. As the robot pose is not included in the global state vector, the robot location estimate is not immediately available on the global level. This is also true when the robot is moving in a previously explored area, as the robot never

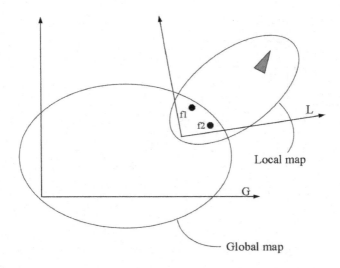

Fig. 4.6   Localization in the global map.

enters an old local map. However, at any time, the global robot location estimate can be easily obtained using the relationship between the local and global coordinate frames.

As shown in Figure 4.6, the coordinate frames of the global and local maps are denoted as G and L respectively. Suppose features f_1, f_2 are the two previously observed features with the estimates in the global and local maps, $\hat{X}_1^G = [\hat{x}_1^G, \hat{y}_1^G]^T$, $\hat{X}_1^L = [\hat{x}_1^L, \hat{y}_1^L]^T$, $\hat{X}_2^G = [\hat{x}_2^G, \hat{y}_2^G]^T$, $\hat{X}_2^L = [\hat{x}_2^L, \hat{y}_2^L]^T$. The shift and rotation transformation between the two coordinate frames G and L can be computed from $\hat{X}_1^G, \hat{X}_1^L, \hat{X}_2^G, \hat{X}_2^L$. Applying the resulting transformation on the robot pose estimate in the local map, $[\hat{x}_r^L, \hat{y}_r^L, \hat{\phi}_r^L]^T$, the global robot location, $[\hat{x}_r^G, \hat{y}_r^G, \hat{\phi}_r^G]^T$, can be obtained.

4.4.2   Data Association

There are two levels of data association in the proposed SLAM algorithm. The lower level data association is for the traditional EKF SLAM in the local map, and the task is to decide whether a newly observed feature is already included in the local map or not. The upper level data association is for the update of the global map. The task is to decide whether a feature in the local map is already included in the global map, and if yes, find the correspondence.

The lower level data association can be performed either by the Nearest Neighbor method [Dissanayake *et al.* (2001)] or the more robust Joint Compatibility Test [Neira and Tardos (2001)]. It should be pointed out that in the local SLAM process, some previously observed features (which are already in the global map) may be treated as new features in the local map.

For the upper level data association in this algorithm, the robot location is not needed. The two common features between the local map and the existing global map have the same function as the robot pose that can link the two sets of feature estimates geometrically. With the correspondence of the states in the local and global maps about these two common features, the estimate of feature locations in the local map can be transformed into the global map coordinate frame. Thus, the stochastic distance between each feature in the local map and each feature in the global map can be computed. Their correspondences can be determined by the Nearest Neighbor method [Dissanayake *et al.* (2001)] or the Joint Compatibility Test [Neira and Tardos (2001)].

However, if this data association is performed in a naive way, the computational cost will be high as the size of the global map is large. As the local map is built within a region of a limited size, it is not necessary to compare the local features with all the features in the global map. Suppose features f_1^L and f_2^L in the local map are identical with the features f_i^G and f_j^G in the global map. First the maximal distance from the features in the local map to f_1^L, which is denoted as d_{max}^L, is computed. Then in the global map, the distances between each feature to feature f_i^G are checked and only the features whose distances are less than $d_{max}^L + \Delta$ are kept. The Δ is the maximal possible estimation error, which is obtained from feature uncertainties in the global map. The distance $d_{max}^L + \Delta$ should cover all the potentially matched features provided the global map is consistent.

4.4.3 *State and Covariance Recovery*

Since the global map is maintained by an information vector and an information matrix, the global state is not available all the time. For the evaluation of Jacobian in (4.17) as well as the upper level data association, the global state and a submatrix of the covariance matrix need to be recovered from the information vector and the information matrix. The global state and the desired columns of the covariance matrix can be recovered using one of the strategies presented in Section 3.3.

It is worth noting that the global map update and global state recovery only need to be performed when a local map is fused into the global map. They happen at a lower frequency compared with the observation frequency and can be scheduled offline [Williams (2001)].

4.4.4   *When to Start a New Local Map*

The only condition imposed on a local map is that there are at least two common features in any local map and the existing global map and that the correspondence is known. Thus it is sufficient if the local map that is being constructed contains at least two features in the previous local map. This can be easily checked by the data association in the local map. This constraint of two common features can be relaxed by using the robot process information to maintain the relation between two consecutive local maps. An alternative way to relax this constraint could be duplicating the feature observations made at the last robot pose (with a doubled covariance) and fusing one copy into the previous local map and the other into the future local map [Frese and Hirzinger (2001)]. Thus, a new local map can still be started if at least two features (even new ones) are observed.

Certainly, better quality local maps will produce better relative feature location information, thus result in better data association quality on the global level. So it is worth closing small loops to obtain a better quality local map when performing traditional EKF SLAM in the local map.

4.5   Computational Complexity

Let n be the average size of local maps, and k_0 be the total number of local maps. Then the size of the global map will be approximately $N = k_0 n$, if the overlaps among local maps are neglected. In the following, the requirement of storage and computational cost of the algorithm will be discussed with respect to the size of the global and local maps.

4.5.1   *Storage*

At each time, the algorithm needs to maintain the estimate of one local map and one global map.

For the local map, the $O(n)$ dimensional state estimate and the associated covariance matrix of $O(n^2)$ dimension need to be stored. For the global map, the $O(N)$ dimensional information vector and the state vector need to be stored. Although the information matrix is $2N$ by $2N$, only the $O(N)$ non-zero elements need to be stored. $O(n)$ columns of the covariance matrix are also needed.

Therefore, the storage requirement is $O(n^2)$ for the local map and $O(nN)$ for the global map.

4.5.2 Local Map Construction

For each local map, as the number of features involved is n, the computational cost only depends on n. When traditional EKF SLAM is used, the computational cost is $O(n^2)$ in each iteration for building the local map [Guivant and Nebot (2001)].

4.5.3 Global Map Update

The size of the current global map depends on the number of local maps that have been fused into it. Suppose at the moment local map $k + 1$ is required to be fused into the global map. At this time, the approximate size of the global map is kn.

To extract the relative location information from the local map, it is required to compute the covariance matrix of the relative information uncertainty. This includes the evaluation of the Jacobian of the function in (4.11) and the multiplication of the Jacobian with the covariance matrix of the local map estimate. The computational cost of these operations is $O(n^2)$.

For the global map update in (4.17), the size of R_{map} is $2n - 3$ and the Jacobian ∇H_{map} is sparse. The update of the information matrix and the information vector is a simple addition operation. Therefore, the computational cost is $O(n^2)$.

For the upper level data association using the method suggested in Section 4.4.2, it is first required to decide which features in the global map are within the local map region, which requires $O(kn)$ computational effort. Then the features in the local map need to be matched with these features in the global map. The computational cost depends on the particular matching technique that is applied. If the nearest neighbor approach is used, the computational cost for the matching process is $O(n^2)$.

In order to evaluate the Jacobians and perform the upper level data association, the state vector and part of the covariance matrix need to be recovered. As demonstrated in Section 3.7, this cost will be of $O(kn^2)$ in most scenarios.

4.5.4 Rescheduling the Computational Effort

Note that the update of the global map only needs to be carried out when a local map is finished and ready to be fused. Obviously, this operation happens at a much lower frequency as compared to the observation frequency. Thus the computationally intensive fusion of the local map into the global map can be managed in such a way that it will not interfere with the update of the local maps. This is another advantage of the local map strategy [Williams (2001)].

4.6 Computer Simulations

4.6.1 Simulation in a Small Area

A simulation experiment with small number of features was first conducted to evaluate the proposed D-SLAM Local Map Joining Filter. The environment used is a 40 meter square with 196 features arranged in uniformly spaced rows and columns. The robot starts from the left bottom corner of the square and follows a random trajectory, revisiting many features and closing many loops. A sensor with a field of view of 180 degrees and a range of 5 meters is simulated to generate relative range and bearing measurements between the robot and the features.

Figures 4.7(a) and 4.7(b) show two of the five local maps. Each local map is created by traditional EKF SLAM. The coordinate frame of the local map is decided by the robot start position and orientation when starting the local map, and the initial uncertainty of the robot pose is set to zero in the local map estimate. The coordinate frame of the global map is the same as that of local Map 1. Local maps 1 to 5 are fused into the global map in sequence. Local map 1 is fused into the global map directly. For local maps 2 to 5, the relative location information is first extracted, then the information is fused into the existing global map. Figure 4.8 shows the global map generated by fusing all the five local maps using the proposed SLAM algorithm.

It is quite flexible in when to start a local map, the only requirement is there are at least 2 common features between two adjacent local maps – this guarantees that the relative position information with respect to these two common features in the local map can be fused into the global map.

In Figures 4.9(a)-4.9(c), a comparison of the estimate uncertainty of three features obtained using traditional EKF SLAM [Dissanayake *et al.* (2001)], D-SLAM and the D-SLAM Local Map Joining Filter is presented.

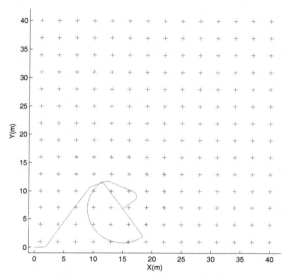

(a) Local Map 1 (features in the local map only include ellipses which indicate the uncertainty of the estimate)

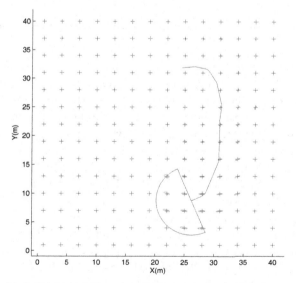

(b) Local Map 5 (features in the local map only include ellipses which indicate the uncertainty of the estimate)

Fig. 4.7 Two local maps.

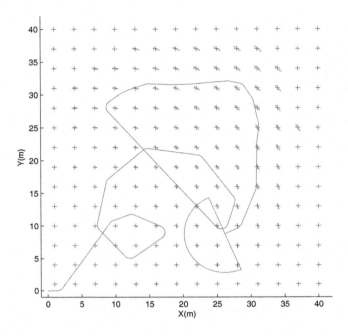

Fig. 4.8   The global map obtained by combining the five local maps.

In Figures 4.9(a)-4.9(c), the solid line is from tradition EKF SLAM; the dotted line is from D-SLAM; the dashed line is from the D-SLAM Local Map Joining Filter. Feature 11 is very near to the robot starting position and is observed from the first local map. It can be seen from Figure 4.9(a) that the results of the three methods are very similar. Figure 4.9(b) shows the effect of the fusion delay. Before the fusion of the third local map (at about loop 1500) the result of the D-SLAM Local Map Joining Filter is worse than D-SLAM because D-SLAM has already fused some of the observation information in the third local map but the results of the D-SLAM Local Map Joining Filter only contain information from the first two local maps. It can be seen that after the fusion of the third local map, the estimate from the D-SLAM Local Map Joining Filter becomes very close to that of traditional EKF SLAM. The fusion delay also leads to the delay of its initialization. Figure 4.9(c) shows that Feature 71, which is far away from the robot starting position, is initialized at loop 1500 (when the third local map is fused) in the D-SLAM Local Map Joining Filter. But it was observed at around loop 1100 when it was initialized by the other

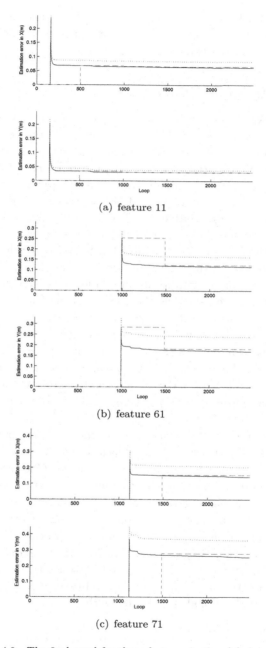

(a) feature 11

(b) feature 61

(c) feature 71

Fig. 4.9 The 2σ bound for three features in the global map.

two SLAM algorithms. It is possible to obtain a conservative estimate of
the global location of features in the local map that is being built using the
global coordinate of the two common features.

4.6.2   *Simulation in a Large Area*

A simulation experiment with a large number of features was conducted to
further evaluate the D-SLAM Local Map Joining Filter. The $105 \times 105m^2$
environment contains 1225 features arranged in uniformly spaced rows and
columns. The robot starts from the left bottom corner of the square and
follows a random trajectory, revisiting many features and closing many
loops. A sensor with a field of view of 180 degrees and a range of 5 meters
is simulated to generate relative range and bearing measurements between
the robot and the features. Figure 4.10(a) shows the robot trajectory and
the feature estimate using traditional EKF SLAM algorithm.

Four hundred small-sized local maps were built using traditional EKF
SLAM. Each local map contains around 10 features. The robot start pose
of one local map is the same as the robot end pose in the previous local
map. Figures 4.11(a), 4.11(b), 4.12(a) and 4.12(b) show four of the local
maps. Figure 4.10(b) shows the global map generated by fusing all the
400 local maps using the D-SLAM Local Map Joining Filter. It can be
seen that the global feature location estimates are consistent since the true
feature locations fall within the 2σ covariance ellipses drawn around the
estimates. The map uncertainty is similar to that obtained by traditional
EKF SLAM.

Figure 4.13(b) shows all the non-zero elements of the sparse information
matrix in black. The matrix is banded as a result of the occasional reordering of the state vector when using the incremental Cholesky Factorization
technique presented in Section 3.3.2.1. Figure 4.13(a) shows the sparse information matrix without any reordering for comparison. This matrix is
obtained in another separate simulation with the same setting by purposely
not applying the reordering during the recovery process.

Figures 4.14(a) and 4.15(a) show the CPU time [1] required to fuse each
local map with and without data association respectively. Figures 4.14(b)
and 4.15(b) show the CPU time required to fuse each local map divided by
the dimension of the global state vector with and without data association

---

[1] All time measurements in this chapter are performed on a desktop with a 2.8GHz Intel
Pentium processor, 1GB of RAM and running Windows. All programs are written in
MATLAB.

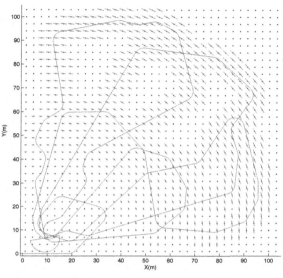

(a) Map from traditional EKF SLAM.

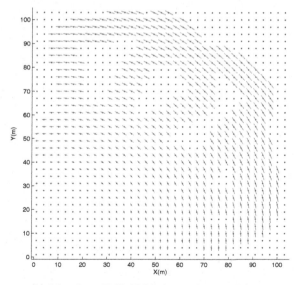

(b) Map from D-SLAM Submap Joining Filter.

Fig. 4.10   Maps from traditional EKF SLAM and D-SLAM Submap Joining Filter.

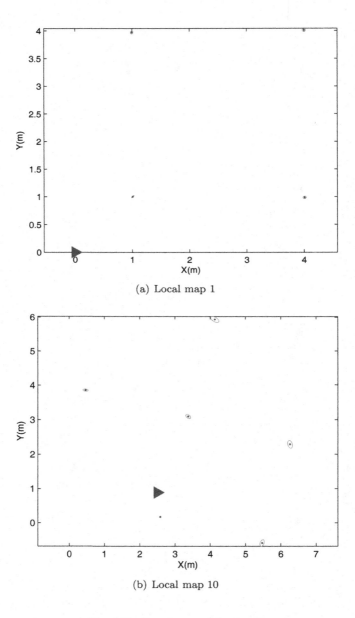

(a) Local map 1

(b) Local map 10

Fig. 4.11   Local maps to be fused.

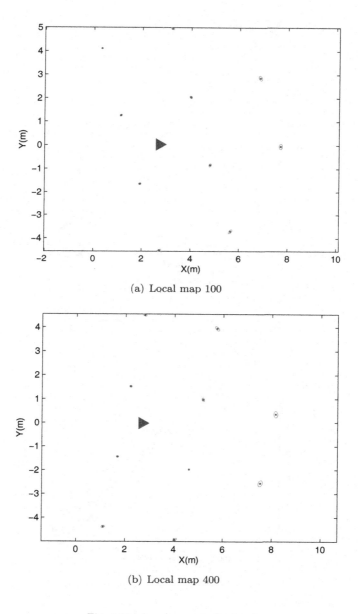

(a) Local map 100

(b) Local map 400

Fig. 4.12   Local maps to be fused.

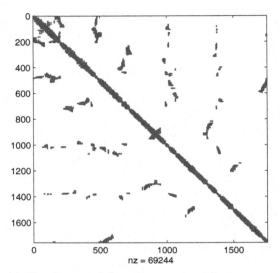

(a) Exactly sparse information matrix without any re-ordering.

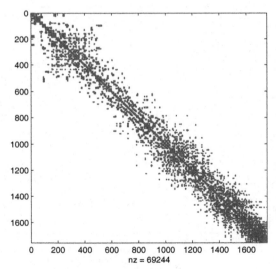

(b) Exactly sparse information matrix with reorder-ing.

Fig. 4.13   Sparse information matrix from D-SLAM Submap Joining Filter.

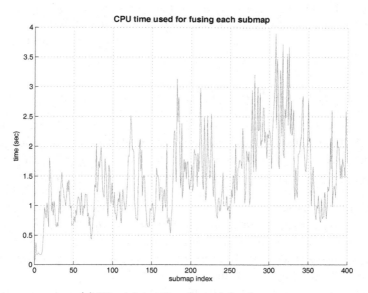

(a) Map joining time for each local map.

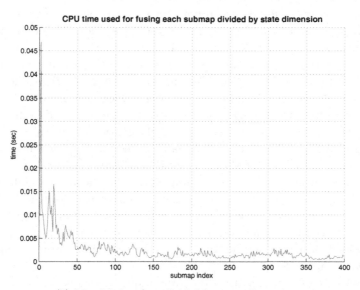

(b) Map joining time divided by state dimension.

Fig. 4.14   Map joining time including data association.

respectively. Figures 4.14(a) and 4.15(a) manifest data association is the major computation in the D-SLAM Local Map Joining Filter, and arguably in almost all SLAM solutions that exploit sparse information matrices.

4.7 Experimental Evaluation

The D-SLAM Local Map Joining Filter was also applied to the popular Victoria Park dataset which was first used in [Guivant and Nebot (2001)]. Neither ground truth nor noise parameters are available for this dataset. Published results for the vehicle trajectory and uncertainty estimates vary, presumably due to different parameters used by various researchers. The results in this section therefore only demonstrate that the D-SLAM Local Map Joining Filter can be applied to this popular dataset.

Figure 4.16(a) presents the vehicle trajectory and the feature location estimates using traditional EKF SLAM. For the results in this section, data association is performed off-line.

The odometry and range-bearing observation data were used to build 200 local maps by traditional EKF SLAM. The robot start pose for one local map is the same as the robot end pose in the previous local map. Figures 4.17(a), 4.17(b), 4.18(a) and 4.18(b) show four of the local maps. In local map 1, as shown in Figure 4.17(a), the vehicle stays still at the starting position for most of the time, therefore the feature uncertainties are very small. In local map 10, shown in Figure 4.17(b), the feature uncertainties are large because of the process noise accumulated due to the vehicle motion. Generally, the quality of the local map depends on the extent of vehicle motion and the rate of observations to features.

Figure 4.16(b) shows the global map obtained by joining 200 local maps using the D-SLAM Local Map Joining Filter. Figure 4.19(b) shows all the non-zero elements of the information matrix in black. The information matrix is not very sparse because the sensor range is relatively large (around $80m$) as compared with the size of the environment ($300m \times 300m$). Figure 4.19(a) shows the information matrix without any reordering. This matrix is obtained from another run of the algorithm by purposely not applying the reordering method.

Figure 4.20(a) shows the CPU time required to fuse each of the 200 local maps. Figure 4.20(b) shows the computation time divided by the dimension of the state vector.

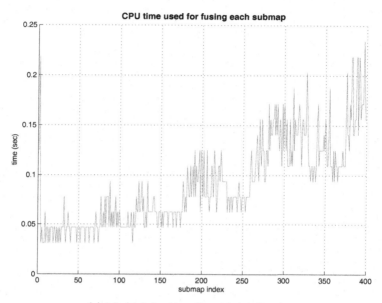

(a) Map joining time for each local map.

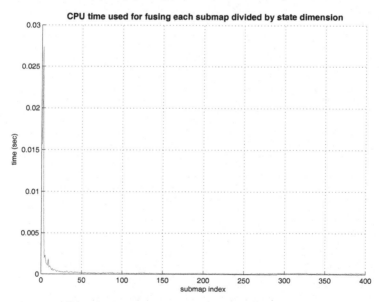

(b) Map joining time divided by state dimension.

Fig. 4.15   Map joining time assuming data association.

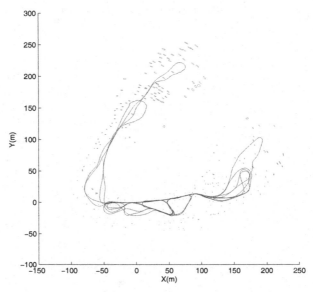

(a) Map obtained by traditional EKF SLAM.

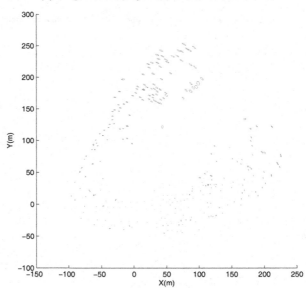

(b) Map obtained by D-SLAM Local Map Joining Filter.

Fig. 4.16   Global maps from experiments using Victoria park dataset.

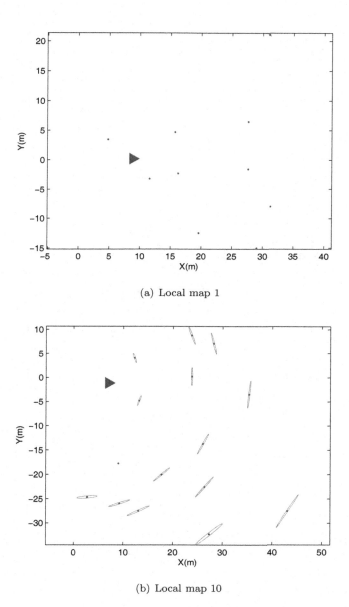

(a) Local map 1

(b) Local map 10

Fig. 4.17   Local maps to be fused.

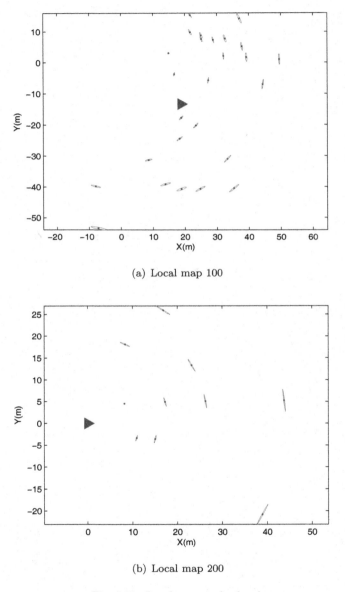

(a) Local map 100

(b) Local map 200

Fig. 4.18   Local maps to be fused.

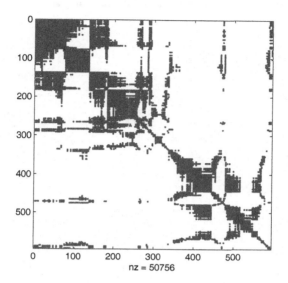

(a) Exactly sparse information matrix without any reordering

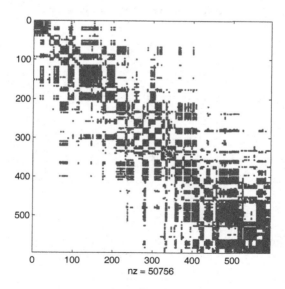

(b) Exactly sparse information matrix with reordering

Fig. 4.19   Exactly sparse information matrix.

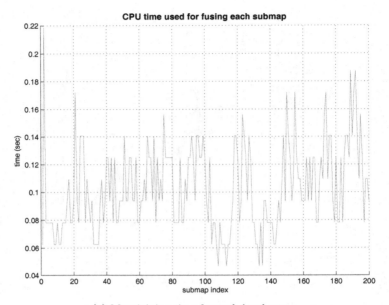

(a) Map joining time for each local map.

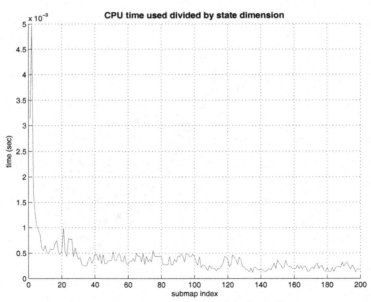

(b) Map joining time divided by state dimension.

Fig. 4.20   Map joining time for experiments using Victoria park dataset.

4.8 Bibliographical Remarks

Submap based algorithms have attracted much attention from the SLAM community in recent years. One approach to SLAM in large environments is to first build local submaps using an EKF-based SLAM algorithm and then combine these local maps into a large scale global map [Tardos *et al.* (2002); Williams (2001)]. The key idea is to build a sequence of local maps which are constructed to be independent of the global map. When fusing a local map into the global map, the local map is first transformed into a common coordinate frame and a joint map state vector is formed. Once the features that appear in both maps are found, a set of constraints relating the appropriate state estimates are used to obtain an updated global map. Algorithms by [Williams (2001)] and [Tardos *et al.* (2002)] follow a similar concept but are developed in a slightly different manner.

In [Tardos *et al.* (2002)], the original state vectors of the local and global maps are separate. Before the fusing operation, state vectors of the two maps are transformed to a common coordinate frame, and then the Joint Compatibility Branch and Bound algorithm [Neira and Tardos (2001)] is used to find the identical features. The constraints of common features are imposed in the form of ideal measurement equations without noise. This local map fusion mechanism is applied in the Local Map Sequencing process to produce a consistent global map at an $O(N^2)$ computational cost, where N is the number of features in the environment.

Williams (2001) proposed the Constrained Local Submap Filter (CLSF), in which the state vector contains the local map, the robot pose in the local map, the global map and the global robot pose when the local map is started. Williams (2001) proved that if a local map is constructed based on a robot pose that is independent of the global map, then the feature location estimates in the local map are independent of those in the global map. Therefore, the local map can be built without affecting the global map. The state estimate of the local map can be transformed into the global coordinate frame using the global robot location stored in the state vector. The transformed local map is then fused into the global map using an estimator that enforces the common feature constraints. The resulting map covariance matrix is fully correlated and thus the map fusion process is computationally demanding. CLSF, however, allows the map fusion to be scheduled off-line while the local map is being built in real-time.

In the Atlas system [Bosse *et al.* (2004)] and the network coupled feature maps (NCFM) [Bailey (2002)], a global map is not maintained at each step.

Instead, the submaps are organized in a global topological graph. Nodes of the graph represent the submaps and the edges represent the transformation or relative location between adjacent submaps. In Atlas, the decision to add a new submap to the graph and close a loop is done using a map-matching algorithm. In the global topological graph, there are multiple paths linking two submaps due to loop closures. The transformation between any two submaps and the associated error estimates are computed following the shortest path in the graph. This is obtained by searching along the edges between adjacent submaps in the graph using the shortest path searching method. When the global map is needed, the transformations from all the submaps to a common coordinate frame are computed using the nonlinear least-squares optimization. By this method, the loop closing constraint is enforced, however in a suboptimal manner. In NCFM, the edges in the global topological graph represent coupling estimates of the relative location between the coordinate origins of two adjacent submaps. The coupling (relative location) between two adjacent submaps is estimated using the constraints of the robot tracks from multiple adjacent submaps and the common features across different submaps. When there are multiple constraints about the coupling, the covariance intersection algorithm [Chen *et al.* (2002)] is used to update the coupling estimates in a consistent manner. The difficult cycle detection problem is solved by an algorithm consisting of two-fold pass detection steps and a confirmation step.

Estrada *et al.* (2005) proposed the Hierarchical SLAM algorithm. Local maps are built using the traditional EKF SLAM algorithm. The local map coordinate origin is selected as the robot start pose, which is identical with the robot end pose of the previous local map. Within a local map, the transformation between the robot start and end poses is maintained. On the global level, a relative stochastic map containing the transformations among the local map coordinate origins is maintained. Nonlinear constrained least-squares optimization is used to impose the loop closure constraints in the global relative map. The global location of each local map can thus be improved when a loop is closed. This algorithm retains the linearization errors in the local area, and reduces the effect of large robot pose estimate uncertainties by using the global nonlinear constrained least-squares optimization. However, the correction effects from the loop closures can not improve the relative location estimates among features within the local map, leading to suboptimal estimates. The same feature may appear in multiple local maps. In these local maps, the estimates of the identical feature are stochastically independent, but they are geometrically dependent. Not enforcing these geometric constraints causes loss of optimality as well and the extent of suboptimality depends on the extent of overlaps between submaps.

4.9 Summary

In this chapter, a SLAM strategy using local maps is presented. The idea is to use traditional EKF SLAM to construct local maps, then extract the relative information of the local feature locations and fuse the information into the global map using the D-SLAM framework.

The algorithm described in this chapter has several properties. (1) Localization is only performed in the SLAM process in building the local map, and the robot pose is not included in the global map (although it can be easily calculated when needed). (2) EIF is used in the global map update. The information matrix associated with the global map is exactly sparse and only the features that are in the same local map are linked through the information matrix. This results in significant savings of computational efforts as compared with the Constrained Local Submap Filter (CLSF) [Williams (2001)] and Local Map Sequencing [Tardos *et al.* (2002)]. (3) The information from the robot process model is exploited in the SLAM process used to build local maps, resulting in more accurate global map than that obtained using D-SLAM described in Chapter 3. (4) The algorithm places very little restriction on how and when local maps are spawned. As the local maps are used to summarize a sequence of observations gathered by the robot, the only requirement is there are at least two previously seen features in each new local map.

In fact, the proposed SLAM algorithm can be regarded as a unifying framework of SLAM algorithms which includes both traditional EKF SLAM and D-SLAM as special cases. When there is only one local map, the algorithm is just the traditional EKF SLAM. When the local map is constructed from a set of observations acquired at one instant of time, the algorithm presented in this chapter is equivalent to D-SLAM.

In contrast to CLSF [Williams (2001)] and Local Map Sequencing [Tardos *et al.* (2002)], the D-SLAM Local Map Joining Filter does not need to transform the features in the local maps into the coordinate frame of the global map. This is due to the fact that the relative location information of features in the local map is extracted and it is invariant with respect to the coordinate frame.

In the D-SLAM Local Map Joining Filter, as the robot location is removed when fusing the local map into the global map, there is some information loss compared to traditional EKF SLAM. It will be shown in the next chapter that this information loss can be avoided by adding a number of robot poses (the robot start pose and the robot end pose in each local map) into the global map state vector.

Chapter 5

Sparse Local Submap Joining Filter

This chapter presents a local map joining algorithm for mapping large scale environments, the Sparse Local Submap Joining Filter (SLSJF) [Huang et al. (2008c)].[1] The local maps can be generated by any stochastic mapping algorithms such as the traditional EKF SLAM as long as they produce consistent state estimate and the associated covariance matrix. The resulting local maps, consisting of the estimates of all feature locations and the final robot pose as well as the robot start pose (coordinate origin) are fused using an EIF to form a global representation of the environment.

In the D-SLAM Local Map Joining Filter presented in Chapter 4, relative location information of features in local maps is used to form the global map, resulting in some information loss. In SLSJF, the estimates of feature locations together with the robot start and end poses in the local map are fused. This leads to an exactly sparse information matrix for the global map and results in a SLAM algorithm with no information loss. Furthermore, it will be demonstrated that using robot start and end poses also avoids the constraint that each local map must include at least two features that have been previously seen.

This chapter is organized as follows. In Section 5.1, the structure of the algorithm is stated. Section 5.2 addresses how to fuse local maps into the global map. Section 5.3 discusses why the information matrix of this algorithm is exactly sparse. Section 5.4 addresses implementation issues such as data association and recovery of the global map state and the associated covariances. Simulation results are presented in Section 5.5 and experiment results using practical data are presented in Section 5.6 to evaluate the algorithm. Section 5.7 discusses the computational complexity

---

[1]The MATLAB source code of this algorithm is provided through OpenSLAM (http://openslam.org/).

and information loss. The chapter is summarized in Section 5.8.

5.1   Structure of Sparse Local Submap Joining Filter

The Sparse Local Submap Joining Filter (SLSJF) aims to build a global map by combining a sequence of local maps. SLSJF regards each local map as a "virtual observation" made from the robot start location about local features and a robot pose at which the local map is completed. For example, in Figure 5.1, local map 2 can be regarded as an observation made from the robot start position in local map 2 – robot observes all the features in local map 2 and a robot located at the robot end position. These local maps are fused one by one to form the global map.

On the local level, any SLAM algorithm such as the traditional EKF SLAM can be used to collect local information and summarize it in the local map state estimate. The local map state contains all the feature locations and the robot final pose in the local map. Implicitly, the robot start pose, which is the coordinate origin, is also in the local map state.

On the global level, the local map estimate is fused to form the global map, which contains features and robot poses. For each pair of subsequent local maps, the transformation between the robot end pose in the former local map and the robot start pose in the latter local map represents the

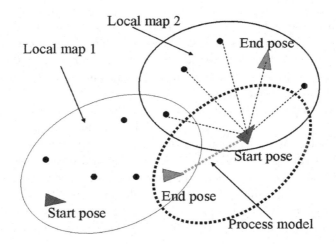

Fig. 5.1   The idea of Sparse Local Submap Joining Filter (SLSJF).

relationship between the two consecutive local maps. For example, the relationship between the two local maps in Figure 5.1 is the relative location between the robot end pose in local map 1 and the robot start pose in local map 2. The transformation can be computed from the robot process model and the information contained is also fused into the global map estimate. This transformation is also used to initialize the robot start pose of the new local map in the global state vector, and thus the correspondence between features in the global map and those in the local map under consideration can be found.

The information available in the local map is only related to a small fraction of the states in the global state vector. It is obvious that the formulation of SLSJF satisfies the conditions set in Section 2.4, therefore results in an exactly sparse information matrix. The sparse structure of the information matrix is similar to that in full SLAM problem [Thrun *et al.* (2005)] where all robot poses and all features are included in the state vector. In SLSJF, however, the number of robot poses included is significantly smaller.

The local map building process can be coordinated such that the robot end pose of the current local map is the same as the robot start pose of the next local map. However, to maintain generality, the case that these two poses are different is used in presenting the algorithm.

5.1.1   *Input to SLSJF - Local Maps*

The input to SLSJF is a sequence of local maps, which are to be constructed by using an estimation-theoretic SLAM algorithm. The information contained in the local map is summarized in the state estimate $\hat{X}^L$ (here the superscript 'L' stands for the local map) and the associated covariance matrix $\mathbf{P}$. The state vector X^L contains the final robot location X_r^L and all the local feature locations $X_1^L, \cdots, X_n^L$ with respect to the local coordinate frame:

$$
X^L = \begin{bmatrix} X_r^L \\ X_1^L \\ \vdots \\ X_n^L \end{bmatrix} = \begin{bmatrix} x_r^L \\ y_r^L \\ \phi_r^L \\ x_1^L \\ y_1^L \\ \vdots \\ x_n^L \\ y_n^L \end{bmatrix}.
\tag{5.1}
$$

The coordinate frame of a local map is defined by the initial robot pose when the building of the local map is started, i.e. the origin and orientation of the local map coordinate frame are decided by the initial robot position and orientation respectively.

The input to SLSJF also includes the relation between each pair of consecutive local maps. The local map k and local map $k + 1$ are related by the robot motion from the robot end pose in local map k to the robot start pose in local map $k + 1$ as described by the process model

$$X_{(k+1)s} = f(X_{ke}, u(k)) + w(k) \qquad (5.2)$$

where $X_{(k+1)s}$ is the robot start pose in local map $k + 1$, X_{ke} is the robot end pose in local map k, $u(k)$ is the control and $w(k)$ is the process noise assumed to be white Gaussian with zero mean and a covariance matrix Q.

The objective of local map joining is to combine the local maps and obtain a global map containing all the features and some robot poses. Similar to the sequential map joining in [Tardos *et al.* (2002)], SLSJF fuses the local maps one by one. At the beginning, the first local map is set as the global map.

5.1.2   *Output of SLSJF - One Global Map*

The state vector of the global map contains feature locations and robot start/end poses in each local map (except the robot start pose of local map 1, which is the origin of the global map). Suppose that the total number of features in the environment is N, and the total number of local maps is p, then the global map state vector is (here the superscript 'G' stands for the global map)

$$X^G = \begin{bmatrix} X_{1e}^G \\ X_{2s}^G \\ X_{2e}^G \\ \vdots \\ X_{ps}^G \\ X_{pe}^G \\ X_1^G \\ \vdots \\ X_N^G \end{bmatrix} \qquad (5.3)$$

where $X_{ks}^G = [x_{ks}^G, y_{ks}^G, \phi_{ks}^G]^T$ is the robot start pose in local map k ($2 \leq k \leq p$), $X_{ke}^G = [x_{ke}^G, y_{ke}^G, \phi_{ke}^G]^T$ is the robot end pose in local map k ($1 \leq k \leq p$), and $X_i^G = [x_i^G, y_i^G]^T$ is the global location of feature i ($1 \leq i \leq N$). Here the subscript 's' stands for 'start pose' and the subscript 'e' stands for 'end pose'.

5.2 Fusing Local Maps into the Global Map

The steps of fusing local map $k + 1$ into the global map using EIF are listed in Algorithm 5.1. The following subsections provide the essential details.

Algorithm 5.1 Fuse local map $k + 1$ into global map using EIF

Require: *global map* and *local map* $k + 1$, robot motion from X_{ke}^G to $X_{(k+1)s}^G$

1: Add $X_{(k+1)s}^G$ into the global map
2: Data association
3: Initialize the new features and $X_{(k+1)e}^G$ in the global map
4: Update the global map
5: Recover the state vector

It is worth noting the time when the recovery of state estimate and the associated covariances should be carried out. State recovery is performed at step 5 in Algorithm 5.1 after the global map is updated at step 4, when the information vector and information matrix have been modified. The covariances recovery is performed during the data association process (step 2 in Algorithm 5.1) described in Section 5.4.1. This is because the part of the covariance matrix that needs to be recovered is decided by the features in the global map that are possible to match the features in the observation. These possibly matched features are to be found in the data association process.

5.2.1 *Adding $X_{(k+1)s}^G$ into the Global Map*

Since the robot end pose in local map k, X_{ke}^G, is already in the global map, the robot start pose in local map $k+1$, $X_{(k+1)s}^G$, can be initialized using the robot motion from X_{ke}^G to $X_{(k+1)s}^G$. This step corresponds to the prediction step in traditional EKF SLAM, although here it is performed as a state augmentation operation and the previous robot pose is not marginalized out.

Suppose the process model is

$$X^G_{(k+1)s} = f(X^G_{ke}, u(k)) + w(k), \tag{5.4}$$

where $u(k)$ is the control and $w(k)$ is the process noise assumed to be white Gaussian with zero mean and a covariance matrix Q.

Denote

$$F(X^G, u(k)) = X^G_{(k+1)s} - f(X^G_{ke}, u(k)), \tag{5.5}$$

then the procedure for adding $X^G_{(k+1)s}$ into the global map is the following.

(1) Add the three elements $\hat{X}^G_{(k+1)s} = f(\hat{X}^G_{ke}, u(k))$ to the state estimate $\hat{X}^G(k)$. Then increase the dimension of $i(k)$ and $I(k)$ by simply adding zero elements corresponding to the new robot pose $X^G_{(k+1)s}$. This results in three new expressions for $\hat{X}^G(k), i(k)$ and $I(k)$;
(2) Update $i(k)$ and $I(k)$ by

$$\begin{aligned} \tilde{I}(k) &= I(k) + \nabla F^T_X Q^{-1} \nabla F_X \\ \tilde{i}(k) &= i(k) + \nabla F^T_X Q^{-1} \nabla F_X \hat{X}^G(k) \end{aligned} \tag{5.6}$$

where ∇F_X is the Jacobian of the function F with respect to X^G evaluated at $\hat{X}^G(k)$.

After the above steps, the robot start pose in local map $k+1$ is initialized in the global map and the robot motion information from X^G_{ke} to $X^G_{(k+1)s}$ is also fused.

5.2.2   *Initializing the Values of New Features and $X^G_{(k+1)e}$ in the Global Map*

In local map $k + 1$, all feature location estimates are with respect to the coordinate origin which is chosen to be the robot start pose. With this robot start pose being initialized as $X^G_{(k+1)s}$ in the global map, all the feature location estimates in local map $k + 1$ can be transformed into the global coordinate frame. Thus, data association can be done to find the correspondences for the old features and identify the new ones. The data association algorithm of SLSJF will be described in Section 5.4.1.

After data association is performed, the initial values of the new features in local map $k+1$ and $X^G_{(k+1)e}$ are computed from the estimated robot start pose $\hat{X}^G_{(k+1)s}$ and the local map estimate $\hat{X}^L$. $\hat{X}^L$ contains the estimates of the final robot pose, $\hat{X}^L_r$, and all the local feature locations, $\hat{X}^L_1, \cdots, \hat{X}^L_n$, with respect to the local coordinate frame.

Suppose the feature f_i in the local map is identified as a new feature by data association, and its local estimate is $\hat{X}^L_i = [\hat{x}^L_i, \hat{y}^L_i]^T$. The initial value for f_i in the global map, $\hat{X}^G_i = [\hat{x}^G_i, \hat{y}^G_i]^T$, can be computed as

$$\hat{x}^G_i = \hat{x}^G_{(k+1)s} + \hat{x}^L_i \cos(\hat{\phi}^G_{(k+1)s}) - \hat{y}^L_i \sin(\hat{\phi}^G_{(k+1)s})$$
$$\hat{y}^G_i = \hat{y}^G_{(k+1)s} + \hat{x}^L_i \sin(\hat{\phi}^G_{(k+1)s}) + \hat{y}^L_i \cos(\hat{\phi}^G_{(k+1)s}), \tag{5.7}$$

where $\hat{x}^G_{(k+1)s}, \hat{y}^G_{(k+1)s}, \hat{\phi}^G_{(k+1)s}$ come from the estimate of the robot start pose in local map $k+1$ as computed in Section 5.2.1. The initial value of $X^G_{(k+1)e}$ can be calculated similarly.

These values are inserted into $\hat{X}^G(k)$ to form a new state vector estimate $\hat{X}^G(k)$. The dimensions of $\tilde{i}(k)$ and $\tilde{I}(k)$ are increased by adding corresponding number of zeros to form a new information vector $\tilde{i}(k)$ and a new information matrix $\tilde{I}(k)$.

Until now, only the initial values of the new feature locations and $X^G_{(k+1)e}$ are put in the state vector. These values are needed to linearize the nonlinear functions relating the local map estimate and the global states which will be used in the next subsection. The information of new features and the robot end pose which is contained in the local map estimate is not fused into the global map yet. It will be presented in the next subsection that all the information from the local map is fused into the global map in one go to update the estimates of the robot poses and feature locations.

5.2.3   *Updating the Global Map*

Suppose local map $k + 1$ is given by (5.1) and the data association result is $X^L_1 \leftrightarrow X^G_{i1}, \cdots, X^L_n \leftrightarrow X^G_{in}$ (including both old and new features). The local map $k + 1$ can be regarded as an observation — a function of the global map state vector (plus noise)

$$z_{map} = \hat{X}^L = H_{map}(X^G) + w_{map} \tag{5.8}$$

where $H_{map}(X^G)$ is given by

$$
\begin{pmatrix}
(x^G_{(k+1)e} - x^G_{(k+1)s})\cos\phi^G_{(k+1)s} + (y^G_{(k+1)e} - y^G_{(k+1)s})\sin\phi^G_{(k+1)s} \\
(y^G_{(k+1)e} - y^G_{(k+1)s})\cos\phi^G_{(k+1)s} - (x^G_{(k+1)e} - x^G_{(k+1)s})\sin\phi^G_{(k+1)s} \\
\phi^G_{(k+1)e} - \phi^G_{(k+1)s} \\
(x^G_{i1} - x^G_{(k+1)s})\cos\phi^G_{(k+1)s} + (y^G_{i1} - y^G_{(k+1)s})\sin\phi^G_{(k+1)s} \\
(y^G_{i1} - y^G_{(k+1)s})\cos\phi^G_{(k+1)s} - (x^G_{i1} - x^G_{(k+1)s})\sin\phi^G_{(k+1)s} \\
\vdots \\
(x^G_{in} - x^G_{(k+1)s})\cos\phi^G_{(k+1)s} + (y^G_{in} - y^G_{(k+1)s})\sin\phi^G_{(k+1)s} \\
(y^G_{in} - y^G_{(k+1)s})\cos\phi^G_{(k+1)s} - (x^G_{in} - x^G_{(k+1)s})\sin\phi^G_{(k+1)s}
\end{pmatrix}
$$

and w_{map} is the noise whose covariance matrix is $\mathbf{P}$.

The formulas of using the observation z_{map} to update the information vector and the information matrix are as follows:

$$I(k+1) = \tilde{I}(k) + \nabla H_{map}^T \mathbf{P}^{-1} \nabla H_{map}$$

$$i(k+1) = \tilde{i}(k) + \nabla H_{map}^T \mathbf{P}^{-1}[z_{map} \qquad (5.9)$$

$$-H_{map}(\hat{X}^G(k)) + \nabla H_{map}\hat{X}^G(k)]$$

where ∇H_{map} is the Jacobian of the function H_{map} with respect to X^G evaluated at $\hat{X}^G(k)$.

5.3   Sparse Information Matrix

All the local maps fused into the global map contain local information and are related to only a small number of the global states. The prediction step between the robot end pose in the previous local map and the robot start pose in the current local map is implemented as a state augmentation operation, without marginalizing out the previous robot pose. So the conditions set in Section 2.4 for the sparseness of the information matrix are satisfied.

Since the function F in (5.5) only relates to the two robot poses X^G_{ke} and $X^G_{(k+1)s}$, its Jacobian with respect to the global state, ∇F_X, is a sparse matrix in the form of

$$\nabla F_X = \left[0, \cdots, 0, \frac{\partial F}{\partial X^G_{ke}}, \frac{\partial F}{\partial X^G_{(k+1)s}}, 0, \cdots, 0\right]. \qquad (5.10)$$

Thus the term $\nabla F_X^T Q^{-1} \nabla F_X$ in (5.6) is an exactly sparse matrix.

Similarly, since $z_{map} = \hat{X}^L$ only involves two robot poses $X^G_{(k+1)s}, X^G_{(k+1)e}$ and a small fraction of all the features in the global map, the matrix $\nabla H^T_{map} \mathbf{P}^{-1} \nabla H_{map}$ in (5.9) is also exactly sparse.

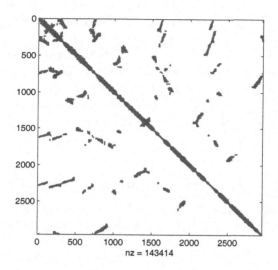

Fig. 5.2   Exactly sparse information matrix from simulation of SLSJF without reordering in Section 5.5.

The sparse information matrix can be illustrated by Figure 5.2 which is the resulting information matrix without reordering in the simulation of SLSJF in Section 5.5. It contains 143414 non-zero elements and 8570890 exactly zero elements.

5.4   Implementation Issues

In fusing the local map into the global map, data association needs to be performed to find the correspondence between the two set of features. The robot start pose in the global map is used to transform the local map estimate into the global coordinate frame.

The recovery of the global map state estimate and a subset of the associated covariance matrix is needed to linearize the nonlinear functions and perform data association.

5.4.1   *Data Association*

The data association algorithm is a probabilistic method assuming that only the geometric relationships among features present in the global and local maps are available.

The features in the current local map can not match the features in a local map which is far away from the current local map coordinate origin in the global map. The local maps which are possible to have overlaps with the current local map are first selected. Data association is only performed about the features in these selected local maps. The current local map is transformed into the coordinate frame of the global map, and then the Joint Compatibility Test with the branch and bound technique [Neira and Tardos (2001)] is performed using the locations of these two sets of features.

5.4.1.1   *Select the Set of Potentially Overlapping Local Maps*

Since the size of local map $k + 1$ is small, it is not necessary to compare the features in local map $k + 1$ with the features in the global map which are far away from $\hat{X}_{(k+1)s}^{G}$.

The local maps that potentially have overlaps with local map $k + 1$ are selected by computing the distance between $\hat{X}_{(k+1)s}^{G}$ and $\hat{X}_{is}^{G}$ $(2 \leq i \leq k)$. If the distance between $\hat{X}_{(k+1)s}^{G}$ and $\hat{X}_{is}^{G}$ is larger than the sum of the two local map radii (the radius of a local map is defined as the maximal distance from the local map features to the origin) plus the possible estimation error, then local map i cannot overlap with local map $k + 1$. For local map 1, the distance between $\hat{X}_{(k+1)s}^{G}$ and $(0, 0, 0)$ is computed.

If local map i is possible to have overlap with local map $k + 1$, then all features in local map i can be found by the links between these features and $\hat{X}_{js}^{G}$ in the information matrix. This is because in the information matrix, only and all the features in local map i have links with $\hat{X}_{js}^{G}$ and $\hat{X}_{je}^{G}$. Only the features in the selected local maps are possible to match the features in local map $k + 1$, and thus are necessary to go through the data association process.

The states corresponding to the candidate features in the global map are available from the state recovery of the previous step. The method to recover the state and the submatrix of the covariance matrix corresponding to the candidate features is described in Section 5.4.2.

5.4.1.2 *Use Joint Compatibility Test to Perform Data Association*

Before comparing the features in local map $k+1$ with the candidate features in the global map, all the feature locations need to be transferred into a common coordinate frame. For example, the features in local map $k+1$ can be transformed into the global coordinate frame. Within local map $k+1$, all the feature positions are relative to the robot start pose, which is $X_{(k+1)s}^{G}$ in the global map. With this relation, the features in local map $k+1$ can be transformed into the global coordinate frame and the covariance can be computed.

Suppose the estimate of feature f_i in local map $k+1$ is $\hat{X}_i^L = [\hat{x}_i^L, \hat{y}_i^L]^T$. The transformed value for f_i in the global map, $\hat{X}_i^G = [\hat{x}_i^G, \hat{y}_i^G]^T$, can be computed as

$$
\begin{aligned}
\hat{x}_i^G &= \hat{x}_{(k+1)s}^G + \hat{x}_i^L \cos(\hat{\phi}_{(k+1)s}^G) - \hat{y}_i^L \sin(\hat{\phi}_{(k+1)s}^G) \\
\hat{y}_i^G &= \hat{y}_{(k+1)s}^G + \hat{x}_i^L \sin(\hat{\phi}_{(k+1)s}^G) + \hat{y}_i^L \cos(\hat{\phi}_{(k+1)s}^G).
\end{aligned}
\tag{5.11}
$$

Put the equations like (5.11) for all the features in local map $k+1$ together and write in the vector form,

$$
\hat{X}_{k+1}^G = S(\hat{X}_{(k+1)s}^G, \hat{X}_{k+1}^L)
\tag{5.12}
$$

where S is the transform function. Suppose the covariance matrix of $\hat{X}_{(k+1)s}^G$ and $\hat{X}_{k+1}^L$ are denoted as $P_{(k+1)s}^G$ and P_{k+1}^L respectively. Since $P_{(k+1)s}^G$ and P_{k+1}^L are independent of each other, the covariance matrix of $\hat{X}_{k+1}^G$, P_{k+1}^G, can be computed as

$$
P_{k+1}^G = \nabla S_{\hat{X}_{(k+1)s}^G} P_{(k+1)s}^G \nabla S_{\hat{X}_{(k+1)s}^G}^T + \nabla S_{\hat{X}_{k+1}^L} P_{k+1}^L \nabla S_{\hat{X}_{k+1}^L}^T
\tag{5.13}
$$

where $\nabla S_{\hat{X}_{(k+1)s}^G}$ is the Jacobian of the function S with respect to $X_{(k+1)s}^G$ evaluated at $\hat{X}_{(k+1)s}^G$, and $\nabla S_{\hat{X}_{k+1}^L}$ is the Jacobian of the function S with respect to X_{k+1}^L evaluated at $\hat{X}_{k+1}^L$.

Since both the number of features in local map $k+1$ and the number of potentially matched features in the global map are limited, the size of the data association problem is small. The robust Joint Compatibility Test with the branch and bound technique [Neira and Tardos (2001)] is used to perform the data association.

5.4.2   *State and Covariance Recovery*

As the global map is maintained by an information vector and an information matrix, the global state estimate $\hat{X}^G$ and the associated covariance matrix are not directly available. For the information vector and information matrix update using (5.6) and (5.9), the state vector needs to be recovered. Also, a submatrix of the covariance matrix (corresponding to the potentially matched global map features) also needs to be recovered for the data association.

The state vector can be recovered by solving the sparse linear equations

$$I(k + 1)\hat{X}^G(k + 1) = i(k + 1). \tag{5.14}$$

This recovery operation is performed after the update of the global map is finished, when the global map state is modified significantly. The complete Cholesky Factorization method described in Section 3.3.2 is used to solve this set of sparse linear equations. The reordering process is performed according to the dimension of H_R defined in equation (3.32) and the threshold. Since the robot start/end poses are in the global map state, the dimension of H_R not only depends on the features in the local map, but also depends on the robot start/end poses.

The desired columns of the covariance matrix, which are decided by the candidate features as described in Section 5.4.1.1, are to be recovered. This can also be obtained by solving a number of sparse linear equations using the complete Cholesky Factorization method described in Section 3.3.2.

5.5   Computer Simulations

A simulation experiment with a large number of features was conducted to evaluate the proposed map joining algorithm. The setting of this simulation is the same as that presented in Section 4.6.2.

The $105 \times 105m^2$ environment contains 1225 features arranged in uniformly spaced rows and columns. The robot starts from the left bottom corner of the square and follows a random trajectory, revisiting many features and closing many loops. A sensor with a field of view of 180 degrees and a range of 5 meters is simulated to generate relative range and bearing measurements between the robot and the features. Figure 5.3 (same as Figure 4.10(a)) shows the robot trajectory and the feature estimate using traditional EKF SLAM algorithm.

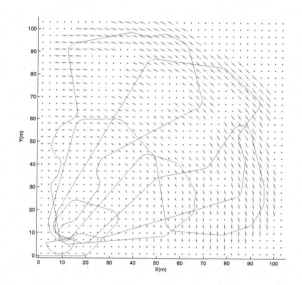

Fig. 5.3   The robot trajectory and map obtained by traditional EKF SLAM.

Four hundred small-sized local maps were built by traditional EKF SLAM. The process of building these local maps is the same as that in the simulation presented in Section 4.6.2. Each local map contains around 10 features and the robot start pose of one local map is the same as the robot end pose in the previous local map. Figures 4.11(a), 4.11(b), 4.12(a) and 4.12(b) show four of the local maps. Figure 5.4 shows the global map generated by fusing all the 400 local maps using SLSJF. It can be seen that the global feature location estimates using SLSJF are consistent since the true feature locations fall within the 2σ covariance ellipses drawn around the estimates. Close examination shows that the map uncertainty is very similar to that obtained by traditional EKF SLAM. Figure 5.5(b) shows the errors and 2σ bounds of the estimates of robot end poses in each local map. It is clear that the estimates are consistent. Comparison with the

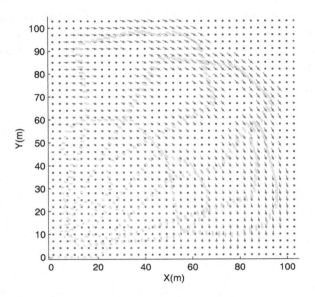

Fig. 5.4   Global map by joining the 400 local maps using SLSJF (shaded ellipses indicate 2σ bounds for the estimates of robot end poses used in local maps).

traditional EKF SLAM results shown in Figure 5.5(a) indicates that the 2σ bounds in SLSJF are smaller. This is expected as the global map state vector in SLSJF contains four hundred robot poses which have an effect similar to that of smoothing [Dellaert and Kaess (2006)].

Figure 5.6(b) shows all the non-zero elements of the sparse information matrix in black. The matrix is banded as a result of the occasional reordering of the state vector when using the complete Cholesky Factorization technique in Section 3.3.2. Figure 5.6(a) shows the sparse information matrix without any reordering for comparison. This matrix is obtained in another separate simulation with the same setting by purposely not applying the reordering during the recovery process.

Figures 5.7(a) and 5.8(a) show the CPU time [2] required to fuse each of the local maps with and without data association respectively. Figures 5.7(b) and 5.8(b) show the CPU time required to fuse each local map divided by the dimension of the global state vector with and without data

---

[2] All time measurements in this chapter are performed on a laptop with a Intel Core 2 Duo processor at 2.2GHz, 3GB of RAM and running Windows, all programs are written in MATLAB.

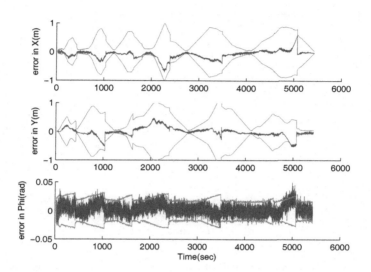

(a) Estimation error of robot pose from EKF

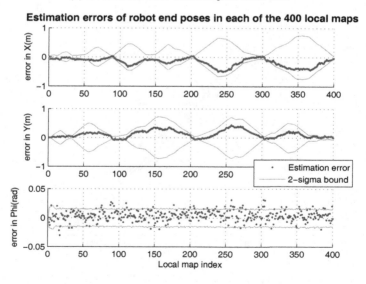

(b) Estimation error of robot end poses in each local map from SLSJF

Fig. 5.5   Estimation error of the robot pose.

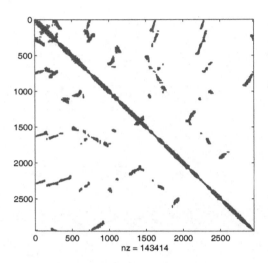

(a) Sparse information matrix obtained by SLSJF without any reordering of the state vector.

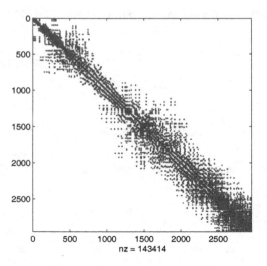

(b) Banded information matrix obtained by SLSJF – it has 143414 non-zero elements and 8570890 exactly zero elements.

Fig. 5.6   Sparse information matrix from SLSJF.

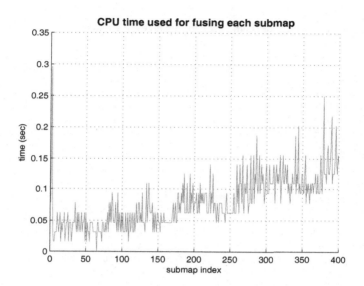

(a) Time used for fusing each local map in SLSJF.

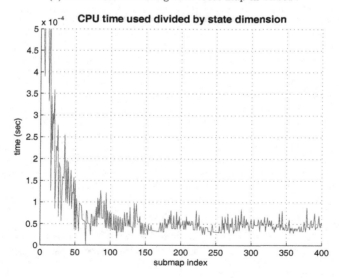

(b) Time used for fusing each local map divided by the state dimension.

Fig. 5.7 Computation time including data association by SLSJF.

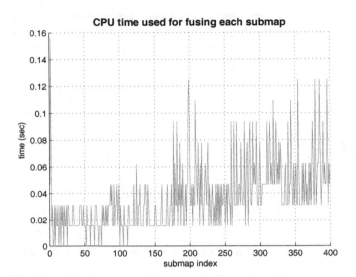

(a) Time used for fusing each local map in SLSJF.

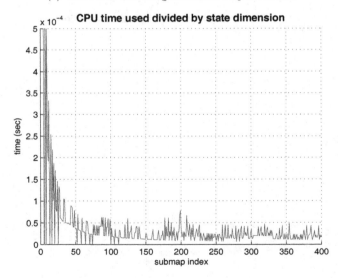

(b) Time used for fusing each local map divided by the state dimension.

Fig. 5.8   Computation time assuming data association by SLSJF.

association respectively. It can be seen from Figures 5.7(a) and 5.8(a) that data association is the major computational cost in SLSJF, and arguably in almost all SLAM solutions that exploit sparse information matrices. In fact, when data association is assumed, it only takes 120 seconds to fuse all the 400 local maps. However, when the data association is not available and needs to be determined using the algorithm described in Section 5.4.1, the time required is 459 seconds.

5.6 Experimental Evaluation

SLSJF was also evaluated with the popular Victoria Park dataset. Neither ground truth nor noise parameters are available for this dataset. The results in this section therefore only demonstrate that SLSJF can be applied to this popular dataset.

Figure 5.9(a) shows the vehicle trajectory and feature location estimates using traditional EKF SLAM. For the results in this section, data association is performed on-line.

The odometry and range-bearing observation data were used to build 200 local maps by EKF SLAM. The robot start pose in one local map is the same as the robot end pose in the previous local map. The process of building these local maps is the same as that in the experiment presented in Section 4.7. Figures 4.17(a), 4.17(b), 4.18(a) and 4.18(b) show four of the local maps.

Figure 5.9(b) presents the global map obtained by joining 200 local maps using SLSJF. Figure 5.10(a) shows all the non-zero elements of the information matrix in black. Figure 5.10(b) shows the information matrix without any reordering. The information matrix is not very sparse because the sensor range is relatively large (around $80m$) as compared with the size of the environment ($300m \times 300m$).

Figure 5.11(a) shows the CPU time to fuse each of the 200 local maps. Figure 5.11(b) shows the computation time divided by the dimension of the global state vector.

5.7 Discussion

5.7.1 *Computational Complexity*

The map joining problem considered in this chapter is similar to that studied in [Tardos *et al.* (2002)] and [Williams (2001)]. The key differences

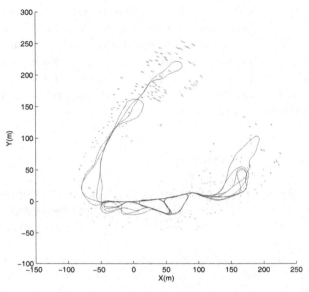

(a) Map from traditional EKF SLAM.

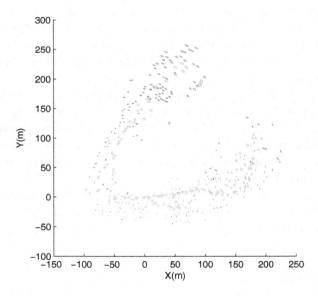

(b) Global map by joining 200 range-bearing local maps using SLSJF (shaded dots indicate robot poses in local maps).

Fig. 5.9   Maps from traditional EKF SLAM and SLSJF.

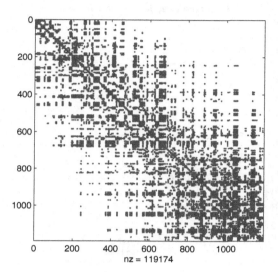

(a) Sparse information matrix with reordering.

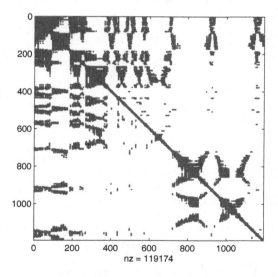

(b) Sparse information matrix without any reordering.

Fig. 5.10 Sparse information matrix from SLSJF.

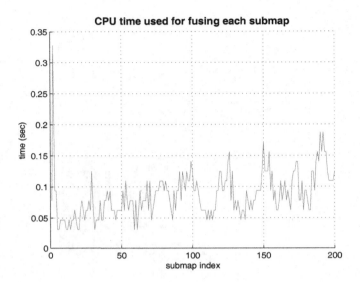

(a) Time used for fusing each of the local maps.

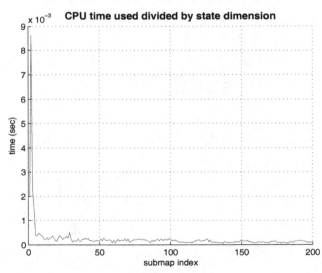

(b) Time used for fusing each of the local maps divided by the state dimension.

Fig. 5.11   Computation time including data association from SLSJF.

between SLSJF and the sequential map joining approach in [Tardos *et al.* (2002)] and the Constrained Local Submap Filter in [Williams (2001)] are: (i) the global state vector used in SLSJF includes not only all the features, but also all the robot start poses and end poses in each local map; (ii) EIF instead of EKF is used in the global map update. The new global state vector together with EIF implementation result in a sparse information matrix and significant computational savings.

When adding the start pose of a local map into the global map, the robot process model is used. In equation (5.6), ∇F_X is extremely sparse with only the elements relating the two robot poses being non-zero. The matrix Q is a 3 by 3 matrix. The computational cost for this step is constant.

When initializing the value for the new features and the robot end pose in the global state vector, the computation operation only relates to the element in the local map as shown in (5.7), so the computation is constant time.

When updating the global map using equation (5.9), the Jacobain ∇H_{map} is exactly sparse with $O(n)$ non-zero elements, where n is the number of features in the local map. The matrix P is a $2n+3$ by $2n+3$ matrix. So the computational cost of (5.9) is $O(n^2)$.

Data association in SLSJF is reduced to a small size problem by first selecting the set of potentially overlapping local maps (Section 5.4.1). By applying the complete Cholesky Factorization procedure in Section 3.3.2, the computational cost of recovering the state vector and a submatrix of the covariance matrix is significantly reduced.

Simulation results show that the computational cost of fusing each local map is around $O(dim)$ where dim is the dimension of the global state vector (Figure 5.7(b)). It is possible that this linear computational cost may not hold for extreme cases if large number of fill-ins occur in the Cholesky Factorization.

5.7.2   *Zero Information Loss*

In SLSJF, all the information in the local maps are used when fusing them into the global map. During the global map update, the information in one local map improves the estimates of all the feature locations and robot poses in the global map. The process information from X_{ke}^G to $X_{(k+1)s}^G$ is also exploited, as these robot poses are estimated and maintained in the global state. Thus, there is no information loss in the local map joining process. This is different from ESEIF [Walter *et al.* (2007)], the D-SLAM algorithm presented in Chapter 3 and the D-SLAM Local Map Joining Filter presented in Chapter 4 where the sparse representation is achieved at the cost of some information loss. Results presented in Section 5.5 demonstrate that the uncertainty of the feature location estimates is almost the same as that of traditional EKF SLAM. It can be argued that the small differences in the feature covariances are due to the effect of difference in linearization. Furthermore, the uncertainty of the robot location estimates is in fact smaller than that produced by traditional EKF SLAM due to the presence of smoothing effect.

5.7.3   *Tradeoffs in Achieving Exactly Sparse Representation*

A number of 2D simulations were conducted to analyze the tradeoff among the D-SLAM algorithm presented in Chapter 3, the D-SLAM Local Map Joining Filter presented in Chapter 4, and the SLSJF in this chapter. The three filters were applied to a dataset, where the robot made 112894 observations to 876 features from 18005 locations and closed many loops. The robot trajectory and the environment can be seen in Figure 5.3. As a comparison, the approach of applying an EIF to the full SLAM formulation was also evaluated using the same dataset.

The simulation results relating performances of the filters in terms of sparseness are summarized in Table 5.1. These results include the number of non-zeros in the information matrix, the number of features/poses in the state vector as well as the state vector dimension.

Table 5.1   Performance of four exactly sparse information filters

Filter	Number of Non-zeros	Number of features	Number of poses	State dimension
full SLAM	1840482	876	18004	55764
D-SLAM	54296	876	0	1752
D-SLAM local map joining filter	69244	876	0	1752
SLSJF	143414	876	400	2952

Regarding the computational savings by exploiting its sparseness the number of non-zeros in the information matrix is the most important. As can be seen from Table 5.1, in all the three filters proposed in this book, the number of non-zeros is significantly reduced compared with full SLAM.

This is a direct result of the fact that in the dataset the number of poses is much larger than the number of features, which is not uncommon in practice.

Close examinations show that most of the non-zeros in the information matrix of the full SLAM approach correspond to pose-pose links and pose-feature links. Therefore containing all robot poses in the full SLAM state vector accounts for the overwhelming lead in the number of non-zeros over other methods. Whereas in SLSJF, only a few poses are retained in the state vector, and thus a very low number of non-zeros is achieved compared with full SLAM, while all the information from the observations and the process model is maintained. In these two approaches, the number of non-zeros is unbounded regarding the size of the environment. In other words, if the robot keeps moving in the same area, the number of non-zeros will keep increasing.

In D-SLAM and D-SLAM Local Map Joining Filter, the number of non-zeros in the information matrix is bounded, regardless of the number of robot poses. In D-SLAM, the bound is related to the feature density and the sensor range; while in D-SLAM Local Map Joining Filter, the bound is related to the feature density and the size of each local map. More non-zeros are observed in D-SLAM Local Map Joining Filter because the size of the local map is larger than the sensor range. Given the size of the environment, the bounded number of non-zeros in these two approaches is achieved at the expense of some information loss.

5.8 Summary

By treating the local map as a virtual observation made from the robot start pose of the local map, a sparse information filter for local map joining, SLSJF, is developed.

By adding robot start poses and end poses of the local maps into the global state vector, information loss in the local map fusion process is completely avoided. The number of robot poses which are added is significantly less than the number of robot poses in the whole robot trajectory. When the size of the local maps is significantly smaller than the size of the global map, the information matrix in SLSJF is very sparse which leads to significant computational savings. Simulation results show that the computational cost for local map fusion is $O(dim)$, where dim is the dimension of the global state vector.

In contrast to the two level mapping algorithms (e.g. [Estrada *et al.* (2005); Bosse *et al.* (2004); Bailey (2002)]), where a set of local maps is organized in a multiple level structure and their relative locations are maintained, SLSJF explicitly estimates a global stochastic map. The local maps are fused into the global map in such a manner that all the local information is exploited, resulting in an optimal local map joining solution.

For the successful application of local map joining using SLSJF, it is important that all the local maps are consistent, because inconsistency of local maps will result in wrong data association of features in the local maps and the global map. Thus, it is essential to use a reliable SLAM algorithm to build the local maps.

Appendix A

Proofs of EKF SLAM Convergence and Consistency

A.1 Matrix Inversion Lemma

The following matrix inversion lemma is used frequently in the proofs of the results in this paper. It can be found in many textbooks about matrices or Kalman Filter (e.g. [Zhang (1999)]).

Lemma A.1. *[Zhang (1999)] Suppose that the partitioned matrix*

$$M = \begin{bmatrix} A & B \\ C & D \end{bmatrix}$$

is invertible and that the inverse is conformably partitioned as

$$M^{-1} = \begin{bmatrix} X & Y \\ U & V \end{bmatrix}, \tag{A.1}$$

where A, D, X and V are square matrices. If A is invertible, then

$$\begin{aligned}
X &= A^{-1} + A^{-1}B(D - CA^{-1}B)^{-1}CA^{-1}, \\
Y &= -A^{-1}B(D - CA^{-1}B)^{-1}, \\
U &= -(D - CA^{-1}B)^{-1}CA^{-1}, \\
V &= (D - CA^{-1}B)^{-1}.
\end{aligned} \tag{A.2}$$

If D is invertible, then

$$\begin{aligned}
X &= (A - BD^{-1}C)^{-1}, \\
Y &= -(A - BD^{-1}C)^{-1}BD^{-1}, \\
U &= -D^{-1}C(A - BD^{-1}C)^{-1}, \\
V &= D^{-1} + D^{-1}C(A - BD^{-1}C)^{-1}BD^{-1}.
\end{aligned} \tag{A.3}$$

Thus if both A and D are invertible,

$$(A - BD^{-1}C)^{-1} = A^{-1} + A^{-1}B(D - CA^{-1}B)^{-1}CA^{-1}. \tag{A.4}$$

When $B = C^T$, equation (A.4) can be written as (substituting D by $-D$)

$$(A + C^T D^{-1} C)^{-1} = A^{-1} - A^{-1}C^T(D + CA^{-1}C^T)^{-1}CA^{-1}. \tag{A.5}$$

A.2   Proofs of EKF SLAM Convergence

Proof of Theorem 1.5: Since the observation noise covariance matrix is R_A, the information gain from one observation is (see (1.66)):

$$I_{new} = \nabla H_A^T R_A^{-1} \nabla H_A. \tag{A.6}$$

For convenience, denote

$$H_A = [e \ \ A]. \tag{A.7}$$

Thus

$$\nabla H_A = [-H_A \ \ A].$$

The total information after the n observations is (see the second equation in (1.65))

$$
\begin{aligned}
I_{A_{end}}^n &= \begin{bmatrix} I_0 & 0 \\ 0 & 0 \end{bmatrix} + n \begin{bmatrix} -H_A^T \\ A^T \end{bmatrix} R_A^{-1} \begin{bmatrix} -H_A & A \end{bmatrix} \\
&= \begin{bmatrix} I_0 + n H_A^T R_A^{-1} H_A & -n H_A^T R_A^{-1} A \\ -n A^T R_A^{-1} H_A & n A^T R_A^{-1} A \end{bmatrix}.
\end{aligned} \tag{A.8}
$$

By the matrix inversion lemma (equations (A.1),(A.3) in Lemma A.1 in Appendix A.1)

$$
\begin{aligned}
P_{A_{end}}^n &= (I_{A_{end}}^n)^{-1} \\
&= \begin{bmatrix} I_0^{-1} & I_0^{-1} H_A^T A^{-T} \\ A^{-1} H_A I_0^{-1} & P_{A_m}^n \end{bmatrix} \\
&= \begin{bmatrix} P_0 & P_0 H_A^T A^{-T} \\ A^{-1} H_A P_0 & P_{A_m}^n \end{bmatrix}
\end{aligned} \tag{A.9}
$$

where

$$P_{A_m}^n = A^{-1} H_A P_0 H_A^T A^{-T} + \frac{A^{-1} R_A A^{-T}}{n}. \tag{A.10}$$

Equation (A.9) is the same as equation (1.79) because

$$A^{-1} H_A = [A^{-1} e \ \ E] = A_e. \tag{A.11}$$

When $n \to \infty$, the second item in (A.10) goes to 0, so (1.80) holds. The proof is completed.

Proof of Lemma 1.1: Since the robot moves from **A** to **B** following the process model, the Jacobians ∇H_A and ∇H_B are not independent. By (1.76),

$$A^{-1} = \begin{bmatrix} \frac{dx_A}{r_A} & -dy_A \\ \frac{dy_A}{r_A} & dx_A \end{bmatrix}, \quad A^{-1} e = \begin{bmatrix} -dy_A \\ dx_A \end{bmatrix}. \tag{A.12}$$

Similarly,

$$B^{-1}e = \begin{bmatrix} -dy_B \\ dx_B \end{bmatrix}.$$

Note that the relationship between the positions of point **A** and point **B** is:

$$\begin{aligned} x_B &= x_A + vT\cos(\phi_A) \\ y_B &= y_A + vT\sin(\phi_A). \end{aligned} \qquad (A.13)$$

Thus

$$\begin{aligned} dx_B &= x_m - x_B = dx_A - vT\cos(\phi_A); \\ dy_B &= y_m - y_B = dy_A - vT\sin(\phi_A). \end{aligned}$$

So

$$A^{-1}e - B^{-1}e = \begin{bmatrix} -vT\sin(\phi_A) \\ vT\cos(\phi_A) \end{bmatrix}.$$

From (1.61),

$$\begin{aligned} B_e \nabla f^A_{\phi X_r} &= [B^{-1}e \quad E] \begin{bmatrix} 1 & 0 \\ \begin{bmatrix} -vT\sin(\phi_A) \\ vT\cos(\phi_A) \end{bmatrix} & E \end{bmatrix} \\ &= \begin{bmatrix} B^{-1}e + \begin{bmatrix} -vT\sin(\phi_A) \\ vT\cos(\phi_A) \end{bmatrix} & E \end{bmatrix} \\ &= [A^{-1}e \quad E] \\ &= A_e. \end{aligned}$$

The proof of this lemma is completed.

Proof of Theorem 1.7: Suppose the robot observed n times ($n \to \infty$ will be considered later) the landmark m at point **A**. Before the robot moves to point **B**, the covariance matrix is $P^n_{A_{end}}$ given by (1.79). By the prediction formula (1.59), the covariance matrix when the robot reaches point **B** is

$$P_{B_{start}} = \begin{bmatrix} P_{rr} & P_{rm} \\ P_{mr} & P^n_{A_m} \end{bmatrix} \qquad (A.14)$$

where

$$\begin{aligned} P_{rr} &= \nabla f^A_{\phi X_r} P_0 (\nabla f^A_{\phi X_r})^T + \nabla f^A_{\gamma v} \Sigma (\nabla f^A_{\gamma v})^T \\ P_{rm} &= \nabla f^A_{\phi X_r} P_0 A^T_e \\ P_{mr} &= A_e P_0 (\nabla f^A_{\phi X_r})^T \\ P^n_{A_m} &= A_e P_0 A^T_e + \frac{A^{-1} R_A A^{-T}}{n}. \end{aligned} \qquad (A.15)$$

Similar to (A.7), denote

$$H_B = [e \ \ B].$$ (A.16)

Thus

$$\nabla H_B = [-H_B \ \ B].$$

The total information after l observations at point **B** is

$$I_{B_{end}}^l = I_{B_{start}} + l \begin{bmatrix} -H_B^T \\ B^T \end{bmatrix} R_B^{-1} [-H_B \ \ B]$$ (A.17)

where $I_{B_{start}} = P_{B_{start}}^{-1}$ and R_B is the covariance matrix of the observation noise.

Denote

$$C_B = \nabla H_B = [-H_B \ \ B], \quad D_B = \frac{R_B}{l}.$$ (A.18)

Using the matrix inversion lemma (see (A.5) in Appendix A.1), the covariance matrix after the observations at point **B** is

$$\begin{aligned} P_{B_{end}}^l \\ = (I_{B_{end}}^l)^{-1} \\ = I_{B_{start}}^{-1} - I_{B_{start}}^{-1} C_B^T (D_{CPC})^{-1} C_B I_{B_{start}}^{-1} \\ = P_{B_{start}} - P_{B_{start}} C_B^T (D_{CPC})^{-1} C_B P_{B_{start}} \end{aligned}$$ (A.19)

where

$$D_{CPC} = D_B + C_B P_{B_{start}} C_B^T.$$ (A.20)

By direct computation,

$$C_B P_{B_{start}} = [C_{P1} \ \ C_{P2}]$$ (A.21)

where

$$\begin{aligned} C_{P1} &= \Delta_{AB} P_0 (\nabla f_{\phi X_r}^A)^T - H_B \nabla f_{\gamma v}^A \Sigma (\nabla f_{\gamma v}^A)^T \\ &= \Delta_{AB} P_0 (\nabla f_{\phi X_r}^A)^T - H_{AB} \Sigma (\nabla f_{\gamma v}^A)^T, \\ C_{P2} &= \Delta_{AB} P_0 A_e^T + \frac{1}{n} B A^{-1} R_A A^{-T}, \end{aligned}$$ (A.22)

with

$$\Delta_{AB} = B A_e - H_B \nabla f_{\phi X_r}^A = B (A_e - B_e \nabla f_{\phi X_r}^A)$$ (A.23)

and H_B defined in (A.16) and H_{AB} defined in (1.95).

By Lemma 1.1, $\Delta_{AB} = 0$, so from (A.21),

$$C_B P_{B_{start}} = \begin{bmatrix} -H_{AB} \Sigma (\nabla f_{\gamma v}^A)^T & \frac{1}{n} B A^{-1} R_A A^{-T} \end{bmatrix}.$$ (A.24)

Let $n \to \infty$, then

$$C_B P_{B_{start}} = \begin{bmatrix} -H_{AB}\Sigma(\nabla f_{\gamma v}^A)^T & 0 \end{bmatrix}, \tag{A.25}$$

and

$$D_B + C_B P_{B_{start}} C_B^T = \frac{1}{l}R_B + H_{AB}\Sigma H_{AB}^T. \tag{A.26}$$

So from (A.19), (A.20) and let $n \to \infty$,

$$\begin{aligned} P_{B_{end}}^l &= P_{B_{start}} - \begin{bmatrix} -\nabla f_{\gamma v}^A \Sigma H_{AB}^T \\ 0 \end{bmatrix} (\tfrac{1}{l}R_B + H_{AB}\Sigma H_{AB}^T)^{-1} \\ &\quad \cdot \begin{bmatrix} -H_{AB}\Sigma(\nabla f_{\gamma v}^A)^T & 0 \end{bmatrix} \\ &= P_{B_{start}}^0 + \begin{bmatrix} \Delta_l & 0 \\ 0 & 0 \end{bmatrix} \end{aligned} \tag{A.27}$$

where $P_{B_{start}}^0$ is defined in (1.92) and

$$\begin{aligned} \Delta_l &= \nabla f_{\gamma v}^A \Sigma (\nabla f_{\gamma v}^A)^T - \nabla f_{\gamma v}^A \Sigma H_{AB}^T \\ &\quad \cdot (\tfrac{1}{l}R_B + H_{AB}\Sigma H_{AB}^T)^{-1} H_{AB}\Sigma(\nabla f_{\gamma v}^A)^T \\ &= \nabla f_{\gamma v}^A [\Sigma - \Sigma H_{AB}^T(\tfrac{1}{l}R_B + H_{AB}\Sigma H_{AB}^T)^{-1}H_{AB}\Sigma](\nabla f_{\gamma v}^A)^T. \end{aligned} \tag{A.28}$$

By matrix inversion lemma (equation (A.5) in Appendix A.1),

$$\begin{aligned} &\Sigma - \Sigma H_{AB}^T(\tfrac{1}{l}R_B + H_{AB}\Sigma H_{AB}^T)^{-1}H_{AB}\Sigma \\ &= [\Sigma^{-1} + lH_{AB}^T R_B^{-1} H_{AB}]^{-1} \geq 0. \end{aligned} \tag{A.29}$$

Thus

$$\Delta_l = \nabla f_{\gamma v}^A \Sigma_B^l (\nabla f_{\gamma v}^A)^T \tag{A.30}$$

with Σ_B^l defined in (1.94). By (A.27), (1.91) holds.

It is easy to see from (1.94) that if the matrix $H_{AB}^T R_B^{-1} H_{AB}$ is invertible, then $\Sigma_B^l \to 0$ and hence $P_B^l \to 0$ as $l \to \infty$. The proof is completed.

A.3   Proofs of EKF SLAM Consistency

Proof of Theorem 1.9: The initial robot information is I_0 in (1.74). The final information after the n observations is

$$\begin{aligned} I_1 &= \begin{bmatrix} I_0 & 0 \\ 0 & 0 \end{bmatrix} + \sum_{j=1}^{n} \nabla H_{\tilde{A}_j}^T R_A^{-1} \nabla H_{\tilde{A}_j} \\ &= \begin{bmatrix} i_\phi & b^T & 0 \\ b & I_{xy} & 0 \\ 0 & 0 & 0 \end{bmatrix} + \sum_{j=1}^{n} \begin{bmatrix} -e^T \\ -\tilde{A}_j^T \\ \tilde{A}_j^T \end{bmatrix} R_A^{-1} \begin{bmatrix} -e & -\tilde{A}_j & \tilde{A}_j \end{bmatrix} \\ &= \begin{bmatrix} i_{\phi 1} & b_1^T \\ b_1 & I_{xy1} \end{bmatrix} \end{aligned}$$

where

$$i_{\phi 1} = i_\phi + n e^T R_A^{-1} e,$$

$$b_1 = \begin{bmatrix} b + (\sum_{j=1}^n \tilde{A}_j^T) R_A^{-1} e \\ -(\sum_{j=1}^n \tilde{A}_j^T) R_A^{-1} e \end{bmatrix},$$ (A.31)

$$I_{xy1} = \begin{bmatrix} I_{xy} + I_{\Sigma_j} & -I_{\Sigma_j} \\ -I_{\Sigma_j} & I_{\Sigma_j} \end{bmatrix}$$

with

$$I_{\Sigma_j} = \sum_{j=1}^n \tilde{A}_j^T R_A^{-1} \tilde{A}_j.$$ (A.32)

Since I_{xy} and I_{Σ_j} are all positive definite matrices, it can be proved that

$$I_{xy1}^{-1} = \begin{bmatrix} I_{xy}^{-1} & I_{xy}^{-1} \\ I_{xy}^{-1} & I_{xy}^{-1} + I_{\Sigma_j}^{-1} \end{bmatrix}$$

and hence

$$b_1^T I_{xy1}^{-1} b_1 = b^T I_{xy}^{-1} b + e^T R_A^{-1} (\sum_{j=1}^n \tilde{A}_j) I_{\Sigma_j}^{-1} (\sum_{j=1}^n \tilde{A}_j^T) R_A^{-1} e.$$

Now apply the matrix inversion lemma to I_1,

$$\begin{aligned} P_1 &= I_1^{-1} \\ &= \begin{bmatrix} (i_{\phi 1} - b_1^T I_{xy1}^{-1} b_1)^{-1} & * \\ * & * \end{bmatrix} \\ &= \begin{bmatrix} (i_\phi - b^T I_{xy}^{-1} b + w(n, A))^{-1} & * \\ * & * \end{bmatrix} \end{aligned}$$ (A.33)

where $*$ stands for a matrix that is not cared about, and $w(n, A)$ is defined in (1.112).

By the definition (1.112),

$$w(n, A) = n e^T W e$$ (A.34)

where

$$W = R_A^{-1} - R_A^{-1} (\sum_{j=1}^n \tilde{A}_j)(n \sum_{j=1}^n \tilde{A}_j^T R_A^{-1} \tilde{A}_j)^{-1} (\sum_{j=1}^n \tilde{A}_j^T) R_A^{-1}.$$

Using the inequality

$$n \sum_{j=1}^n \tilde{A}_j^T R_A^{-1} \tilde{A}_j \geq (\sum_{j=1}^n \tilde{A}_j^T) R_A^{-1} (\sum_{j=1}^n \tilde{A}_j),$$ (A.35)

it can be shown that $W \geq 0$ and thus

$$w(n, A) \geq 0.$$

So in (A.33),

$$(i_\phi - b^T I_{xy}^{-1} b + w(n, A))^{-1} \leq (i_\phi - b^T I_{xy}^{-1} b)^{-1} = p_\phi.$$

This means that the updated robot orientation uncertainty cannot be greater than the initial robot orientation uncertainty.

Furthermore, if matrices $\tilde{A}_j, 1 \leq j \leq n$ are all the same, then (A.35) becomes an equality and

$$w(n, A) = 0$$

and hence

$$(i_\phi - b^T I_{xy}^{-1} b + w(n, A))^{-1} = p_\phi.$$

However, if matrices $\tilde{A}_j, 1 \leq j \leq n$ are different, then

$$w(n, A) > 0$$

and

$$(i_\phi - b^T I_{xy}^{-1} b + w(n, A))^{-1} < p_\phi. \tag{A.36}$$

It is obvious that the robot orientation uncertainty cannot be reduced by observing a single new landmark. So this is wrong (inconsistent). In general, if matrices $\tilde{A}_j, 1 \leq j \leq n$ are different, then $w(n, A) \to \infty$ when $n \to \infty$, thus

$$(i_\phi - b^T I_{xy}^{-1} b + w(n, A))^{-1} \to 0.$$

This means that the uncertainty of the robot orientation will decrease to 0 after many observations. The proof is completed.

Proof of Theorem 1.10: The proof is only given for the simple case when there is no control noise, i.e. $\Sigma = 0$. In this case, if $\Delta_{AB} = 0$, then $C_{P1} = 0$ in (A.21); if $\Delta_{AB} \neq 0$, then $C_{P1} = \Delta_{AB} P_0 (\nabla f_{\phi X_r}^A)^T \neq 0$. Now by (A.19) and (A.21), the upper left submatrix of $P_{B_{end}}^l$ is

$$\nabla f_{\phi X_r}^A P_0 (\nabla f_{\phi X_r}^A)^T - C_{P1}^T (D_{CPC})^{-1} C_{P1} \leq \nabla f_{\phi X_r}^A P_0 (\nabla f_{\phi X_r}^A)^T.$$

This violates the lower bound proved in Theorem 1.7.

Appendix B

Incremental Method for Cholesky Factorization of SLAM Information Matrix

In this appendix, it is shown that the Cholesky Factorization of the information matrix $I(k+1)$ can be constructed from that of $I(k)$ due to the similarity between the information matrices of two consecutive steps as shown in (3.15).

B.1 Cholesky Factorization

Suppose the Cholesky Factorization of $I(k)$ is L_k (a lower triangular matrix). Then

$$I(k) = L_k L_k^T. \tag{B.1}$$

Let L_k and $I(k)$ be partitioned according to (3.27) as

$$L_k = \begin{bmatrix} L_{11} & 0 \\ L_{21} & L_{22} \end{bmatrix}, \quad I(k) = \begin{bmatrix} I_{11} & I_{21}^T \\ I_{21} & I_{22} \end{bmatrix}. \tag{B.2}$$

Then from (B.1) and (B.2),

$$
\begin{aligned}
L_{11} L_{11}^T &= I_{11}, \\
L_{21} L_{11}^T &= I_{21}, \\
L_{21} L_{21}^T + L_{22} L_{22}^T &= I_{22}.
\end{aligned}
\tag{B.3}
$$

According to (3.15), (3.27) and (B.2), $I(k+1)$ can be expressed by

$$
\begin{aligned}
I(k+1) &= \begin{bmatrix} I_{11} & I_{21}^T \\ I_{21} & I_{22}^{k+1} \end{bmatrix} \\
&= \begin{bmatrix} I_{11} & I_{21}^T \\ I_{21} & I_{22} + H_R \end{bmatrix}.
\end{aligned}
\tag{B.4}
$$

Lemma B.2. *The Cholesky Factorization of $I(k+1)$ is*

$$L_{k+1} = \begin{bmatrix} L_{11} & 0 \\ L_{21} & L_{22}^{k+1} \end{bmatrix} \tag{B.5}$$

where L_{22}^{k+1} is the Cholesky Factorization of the submatrix $H_R + L_{22}L_{22}^T = I_{22}^{k+1} - L_{21}L_{21}^T$. That is,

$$
\begin{aligned}
L_{22}^{k+1}(L_{22}^{k+1})^T &= H_R + L_{22}L_{22}^T \\
&= I_{22}^{k+1} - L_{21}L_{21}^T.
\end{aligned}
\tag{B.6}
$$

Proof: By (B.3), (B.5) and (B.6),

$$
\begin{aligned}
&L_{k+1}L_{k+1}^T \\
&= \begin{bmatrix} L_{11} & 0 \\ L_{21} & L_{22}^{k+1} \end{bmatrix} \begin{bmatrix} L_{11}^T & L_{21}^T \\ 0 & (L_{22}^{k+1})^T \end{bmatrix} \\
&= \begin{bmatrix} L_{11}L_{11}^T & L_{11}L_{21}^T \\ L_{21}L_{11}^T & L_{21}L_{21}^T + L_{22}^{k+1}(L_{22}^{k+1})^T \end{bmatrix} \\
&= \begin{bmatrix} I_{11} & I_{21}^T \\ I_{21} & L_{21}L_{21}^T + H_R + L_{22}L_{22}^T \end{bmatrix} \\
&= \begin{bmatrix} I_{11} & I_{21}^T \\ I_{21} & I_{22} + H_R \end{bmatrix}.
\end{aligned}
\tag{B.7}
$$

Thus by (B.4),

$$
I(k+1) = L_{k+1}L_{k+1}^T.
\tag{B.8}
$$

Since both L_{11} and L_{22}^{k+1} are lower triangular, L_{k+1} is also lower triangular. Thus L_{k+1} is the Cholesky Factorization of $I(k+1)$.

Remark: It is known that in a Cholesky Factorization an entry of the input matrix affects only the Cholesky factor to the right of that element. This lemma exploits this fact and the special manner that the information matrix evolves to achieve computational savings. When the dimension of H_R in (3.27) is $O(1)$, the computational cost of computing L_{k+1} using (B.5) is $O(1)$, which is much more efficient than directly computing the Cholesky Factorization of $I(k+1)$.

B.2   Approximate Cholesky Factorization

The formula (B.5) in the above section can also be used to incrementally compute an approximate Cholesky Factorization of $I(k+1)$.

Suppose $\tilde{L}_k$ is an approximation of the Cholesky Factorization of $I(k)$, then

$$
I(k) \approx \tilde{L}_k \tilde{L}_k^T.
\tag{B.9}
$$

Let $\tilde{L}_k$ be partitioned according to (3.27) as

$$
\tilde{L}_k = \begin{bmatrix} \tilde{L}_{11} & 0 \\ \tilde{L}_{21} & \tilde{L}_{22} \end{bmatrix}.
\tag{B.10}
$$

Let $\tilde{L}_{22}^{k+1}$ be an approximate Cholesky Factorization of the submatrix $H_R + \tilde{L}_{22}\tilde{L}_{22}^T$ and construct $\tilde{L}_{k+1}$ by

$$\tilde{L}_{k+1} = \begin{bmatrix} \tilde{L}_{11} & 0 \\ \tilde{L}_{21} & \tilde{L}_{22}^{k+1} \end{bmatrix}. \tag{B.11}$$

Then the following approximate equation can be proved in a way similar to the proof of Lemma B.2,

$$I(k+1) \approx \tilde{L}_{k+1}\tilde{L}_{k+1}^T. \tag{B.12}$$

Bibliography

Andrade-Cetto, J., and Sanfeliu, A. 2005. The effects of partial observability when building fully correlated maps. *IEEE Transactions on Robotics*, 21(4):771-777.

Bailey, T. 2002. Mobile robot localization and mapping in extensive outdoor environment. Ph.D. Thesis. Australian Centre of Field Robotics, University of Sydney.

Bailey, T., Nieto, J., Guivant, J., Stevens, M., and Nebot, E. 2006. Consistency of the EKF-SLAM algorithm. In *Proc. IEEE/RSJ International Conference on Intelligent Robots and Systems*, pp. 3562-3568.

Bailey, T., and Durrant-Whyte, H. 2006. Simultaneous localization and mapping (SLAM): part II. *Robotics and Automation Magazine*, 13(3):108-117.

Barrett, R., M. Berry, T. F., and Chan, T. F., Demmel, J., Donato, J., Dongarra, J., Eijkhout, V., Pozo, R., Romine, C., and Vorst, H. 1994. *Templates for the Solution of Linear Systems: Building Blocks for Iterative Methods*. SIAM, Philadelphia.

Bar-Shalom, Y., Li, X. R., and Kirubarajan, T. 2001. *Estimation with Applications to Tracking and Navigation: Theory Algorithms and Software*. John Wiley & Sons.

Borgelt, C., and Kruse, R. 2002. *Graphical Models: Methods for Data Analysis and Mining*. John Wiley & Sons.

Bosse, M., Newman, P. M., Leonard, J. J., and Teller, S. 2004. SLAM in large-scale cyclic environments using the Atlas framework. *International Journal on Robotics Research*, 23(12):1113-1139.

Castellanos, J. A., Neira. J., and Tardos, J. D. 2004. Limits to the consistency of EKF-based SLAM. In *Proc. 2004 IFAC Symposium on Intelligent Autonomous Vehicles*, Lisbon, Portugal.

Chen, L., Arambel, P. O., and Mehra, R. K. 2002. Estimation under unknown correlation: covariance intersection revisited. *IEEE Transactions on Automatic Control*, 47(11):1879-1882.

Csorba, M., Uhlmann, J. K., and Durrant-Whyte, H. 1997. A suboptimal algorithm for automatic map building. In *Proc. American Control Conference*, pp. 537-541.

Deans, M. C., and Hebert, M. 2000. Invariant filtering for simultaneous localization and map building. In *Proc. IEEE International Conference on Robotics and Automation*, pp. 1042-1047.

Dellaert, F. 2005. Square root SAM. In *Proc. Robotics: Science and Systems*.

Dellaert, F., and Kaess, M. 2006. Square Root SAM: simultaneous location and mapping via square root information smoothing. *International Journal of Robotics Research*, 25(12):1181-1203.

Dissanayake, G., Newman, P., Clark, S., Durrant-Whyte, H., and Csobra, M. 2001. A solution to the simultaneous localization and map building (SLAM) problem. *IEEE Transactions on Robotics and Automation*, 17(3):229-241.

Durrant-Whyte, H. F. 1988. Uncertain geometry in robotics. *IEEE Transactions on Robotics and Automation*, 4(1):23-31.

Durrant-Whyte, H., and Bailey, T. 2006. Simultaneous localization and mapping: part I. *Robotics and Automation Magazine*, 13(2):99-110.

Estrada, C., Neira, J., and Tardos J. D. 2005. Hierarchical SLAM: real-time accurate mapping of large environments. *IEEE Transactions on Robotics*, 21(4):588-596.

Eustice, R. M., Singh, H., and Leonard, J. 2005a. Exactly sparse delayed-state filters. In *Proc. IEEE International Conference on Robotics and Automation*, pp. 2428-2435.

Eustice, R. M., Walter, M., and Leonard, J. 2005b. Sparse extended information filters: insights into sparsification. In *Proc. IEEE/RSJ International Conference on Intelligent Robots and Systems*, pp. 641-648.

Eustice, R. M., Singh, H., Leonard, J., Walter, M., and Ballard, R. 2005c. Visually navigating the RMS Titanic with SLAM information filters. In *Proc. Robotics: Science and Systems*.

Eustice, R. M., Singh, H., and Leonard, J. 2006. Exactly sparse delayed-state filters for view-based SLAM. *IEEE Transactions on Robotics*, 22(6):1100-1114.

Feder, H., Leonard, J., and Smith, C. 1999. Adaptive mobile robot navigation and mapping. *International Journal of Robotics Research*, 18(7):650-668.

Fenwick, J. W., Newman, P. M., and Leonard, J. J. 2002. Cooperative concurrent mapping and localization. In *Proc. IEEE International Conference on Robotics and Automation*, pp:1810-1817.

Folkesson, J., and Christensen, H. I. 2004. Graphical SLAM - a self-correcting map. In *Proc. IEEE International Conference on Robotics and Automation*, pp. 383-390.

Frese, U., and Hirzinger, G. 2001. Simultaneous localization and mapping – a discussion. In *Proc. IJCAI Workshop on Reasoning with Uncertainty in Robotics*.

Frese, U. 2004. An O(log n) algorithm for simulateneous localization and mapping of mobile robots in indoor environments. Ph.D. Thesis. University of Erlangen-Nürnberg.

Frese, U. 2005. A proof for the approximate sparsity of SLAM information matrices. In *Proc. IEEE International Conference on Robotics and Automation*, pp.331-337.

Frese, U., Larsson, P., and Duckett, T. 2005. A multilevel relaxation algorithm for simultaneous localization and mapping. *IEEE Transactions on Robotics* 21(2):196-207.

Frese, U. 2006a. A discussion of simultaneous localization and mapping. *Autonomous Robots* 20 (1):25-42.

Frese, U. 2006b. Treemap: An O(log n) algorithm for indoor simultaneous localization and mapping. *Autonomus Robots*, 21:103-122.

Gerkey, B. P., Vaughan, R. T., and Howard, A. 2003. The player/stage project: Tools for multirobot and distributed sensor systems. In *Proc. International Conference on Advanced Robotics*, pp. 317-323.

Guivant, J. E., and Nebot, E. M. 2001. Optimization of the simultaneous localization and map building (SLAM) algorithm for real time implementation. *IEEE Transactions on Robotics and Automation*, 17(3):242-257.

Huang, S., Wang, Z., and Dissanayake, G. 2006. Mapping large scale environments using relative position information among landmarks. In *Proc. International Conference on Robotics and Automation*, pp. 2297-2302.

Huang S., and Dissanayake G. 2007. Convergence and consistency analysis for Extended Kalman Filter based SLAM. *IEEE Transactions on Robotics*, 23(5):1036-1049.

Huang, S., Wang, Z., and Dissanayake, G. 2008a. Exact state and covariance submatrix recovery for submap based sparse EIF SLAM algorithms. In *Proc. IEEE International Conference on Robotics and Automation*, pp.1868-1873.

Huang, S., Wang, Z., Dissanayake, G., and Frese, U. 2008b. Iterated SLSJF: A sparse local submap joining algorithm with improved consistency, In *Proc. Australasian Conference on Robotics and Automation*.

Huang, S., Wang, Z., and Dissanayake, G. 2008c. Sparse local submap joining filter for building large scale maps. *IEEE Transactions on Robotics*, 24(5):1121-1130.

Huang, S., Wang, Z., Dissanayake, G., and Frese, U. 2009. Iterated D-SLAM Map Joining: evaluating its performance in terms of consistency and efficiency. *Autonomous Robots, Special Issue on Characterizing Mobile Robot Localization and Mapping*, 27(4):409-429.

Julier, S. J., and Uhlmann, J. K. 2001a. Simultaneous localization and map building using split covariance intersection. In *Proc. IEEE/RSJ International Conference on Intelligent Robots and Systems*, pp. 1257-1262.

Julier, S. J., and Uhlmann, J. K. 2001b. A counter example for the theory of simultaneous localization and map building. In *Proc. IEEE International Conference on Robotics and Automation*, pp. 4238-4243.

Kaess, M., Ranganathan, A., and Dellaert, F. 2007. Fast incremental square root information smoothing, In *Proc.International Joint Conferences on Artificial Intelligence (IJCAI)*, pp. 2129-2134.

Kim, S. J. 2004. Efficient simultaneous localization and mapping algorithms using submap networks. Ph.D. Thesis. Massachusetts Institute of Technology.

Krauthausen, P., Kipp, A., and Dellaert, F. 2006. Exploiting locality in SLAM by nested dissection, In *Proc. Robotics: Science and Systems*.

Lauritzen, S. L. 1996. *Graphical Models*. Oxford University Press.

Lee, K. W., Wijesoma, W. S., and Javier, I. G. 2006. On the observability and observability analysis of SLAM. In *Proc. IEEE/RSJ, International Conference on Intelligent Robots and Systems*, pp:3569-3574.

Lowe D. G. 2004. Distinctive image features from scale-invariant keypoints. *International Journal of Computer Vision*, 60(2):91-110.

Manyika, J., and Durrant-Whyte, H. 1994. *Data Fusion and Sensor Management: A Decentralized Information-theoretic Approach*. Ellis Horwood, New York.

Martinelli, A., Tomatis, N., and Siegwart, R. 2004. Open challenges in SLAM: An optimal solution based on shift and rotation invariants. In *Proc. IEEE International Conference on Robotics and Automation*, pp. 1327-1332.

Martinelli, A., Tomatis, N., and Siegwart, R. 2005. Some results on SLAM and the closing the loop problem. In *Proc. IEEE/RSJ, International Conference on Intelligent Robots and Systems*, pp:334-339.

Maybeck, P. 1979. *Stochastic Models, Estimation, and Control. Vol.1.* Academic Press, New York.

Montemerlo, M., Thrun, S., Koller, D., and Wegbreit, B. 2002. Fast-SLAM: A factored solution to the simultaneous localization and mapping problem. *AAAI National Conference on Artificial Intelligence*, pp. 593-598.

Moutarlier, P., and Chatlia, R. 1989a. Stochastic multisensor data fusion for mobile robot localization and environment modeling. In *Proc. International Symposium on Robotics Research*, pp. 85-94.

Moutarlier, P., and Chatila, R. 1989b. An experimental system for incremental environment modeling by an autonomous mobile robot. In *Proc. International Symposium on Experimental Robotics*.

Neira, J., and Tardos, J. D. 2001. Data association in stochastic mapping using the joint compatibility test. *IEEE Transactions Robotics and Automation*, 17(6):890-897.

Nettleton, E. W., Gibbens, P. W., and Durrant-Whyte, H. F. 2000a. Closed form solutions to the multiple platform simultaneous localization and map building (slam) problem. In *Sensor Fusion: Architectures, Algorithms, and Applications IV* (eds Dasarathy, B. V.), Vol. 4051, pp. 428-437, Bellingham, WA.

Nettleton, E., Durrant-Whyte, H., Gibbens, P., and Goktogan, A. 2000b. Multiple platform localization and map building. In *Sensor Fusion and Decentralized Control in Robotic Systems III* (eds McKee, G. T., and Schenker, P.S.), Vol. 4196, pp. 337-347, Bellingham, WA.

Newman, P. 2000. On the structure and solution of the simultaneous localization and map building problem. Ph.D. Thesis. Australian Centre of Field Robotics, University of Sydney.

Paskin, M. 2003. Thin junction tree filters for simultaneous localization and mapping. In *Proc.International Joint Conferences on Artificial Intelligence (IJCAI)*, pp. 1157-1164.

Pissanetzky, S. 1984. *Sparse Matrix Technology.* Academic Press, London.

Pradalier, C., and Sekhavat, S. 2003. Simultaneous localization and mapping using the Geometric Projection Filter and correspondence graph matching. *Advanced Robotics* 17(7):675-690.

Saad, Y. 1996. *Iterative Methods for Sparse Linear Systems.* PWS Publishing Company. (Electronic Version)

Salmond, D. J., and Gordon, N. J. 2000. Group tracking with limited sensor resolution and finite field of view. In *Proc. of SPIE*, 4048: 532-540.

Shewchuk, J. 1994. An Introduction to the Conjugate Gradient method without the agonizing pain. Technical Report. CMU-CS-94-125, Carnegie Mellon Univerisity, Pittsburgh, PA, USA.

Siciliano, B., Khatib, O. (Eds.) 2008. *Springer Handbook of Robotics.* Springer, Berlin Heidelberg.

Smith, R., Self, M., Cheeseman, P. 1987. A stochastic map for uncertain spatial relationships. In *Proc. International Symposium on Robotics Research*, pp. 467-474.

Smith, R., Self, M., and Cheeseman, P. 1990. Estimating uncertain spatial relationships in robotics. In *Automomous Robot Vehicles*, Cox, I. J., and Wilfon, G. T., eds, New York: Springer Verleg. pp. 167-193.

Speed, T. P., and Kiiveri, H. T. 1986. Gaussian Markov distributions over finite graphs. *The Annals of Statistics*, 14(1):138-150.

Tardos, J. D., Neira, J., Newman, P., and Leonard, J. 2002. Robust mapping and localization in indoor environments using sonar data. *International Journal of Robotics Research*, 21(4):311-330.

Thrun, S., Liu, Y., Koller, D., Ng, A. Y., Ghahramani, Z., and Durrant-Whyte, H. 2004a. Simultaneous localization and mapping with sparse extended information filters. *International Journal of Robotics Research*, 23(7-8):693-716.

Thrun, S., Montemerlo, M., Koller, D., Wegbreit, B., Nieto, J., and Nebot, E. 2004b. FastSLAM: an efficient solution to the simultaneous localization and mapping problem with unknown data association. *Journal of Machine Learning Research.*

Thrun, S. and Montemerlo, M. 2004. The GraphSLAM algorithm with applications to large-scale mapping of urban structures. *International Journal of Robotics Research*, 25(5-6):403-429.

Thrun, S., Burgard, W., and Fox. D. 2005. *Probabilistic Robotics.* The MIT Press.

Vidal-Calleja, T., Bryson, M., Sukkarieh, S., Sanfeliu, A., and Andrade-Cetto, J. 2007. On the observability of bearing-only SLAM. In *Proc. IEEE International Conference on Robotics and Automation*, pp:4114-4119.

Walter, M., Eustice, R., and Leonard, J. 2005. A provably consistent method for imposing exact sparsity in feature-based SLAM information filters. In *Proc. International Symposium on Robotics Research.*

Walter, M., Eustice, R., and Leonard, J. 2007. Exactly sparse extended information filters for feature-based SLAM. *International Journal of Robotics Research*, 26(4): 335-359.

Wang, Z., Huang, S., and Dissanayake, G. 2005. Decoupling localization and mapping in SLAM using compact relative maps. In *Proc. IEEE/RSJ International Conference on Intelligent Robots and Systems*, pp. 3336-3341.

Wang, Z., Huang, S., and Dissanayake, G. 2007a. D-SLAM: a decoupled solution to simultaneous localization and mapping. *International Journal of Robotics Research*, 26(2): 187-204.

Wang, Z., Huang, S., and Dissanayake, G. 2007b. Tradeoffs in SLAM with sparse information filters. *International Conference on Field and Service Robotics*, pp. 253-262.

Wang, Z. 2007. Exactly sparse information filters for simultaneous localization and mapping. Ph.D. Thesis. University of Techology, Sydney.

Wang, Z., and Dissanayake, G. 2008. Observability analysis of SLAM using Fisher Information Matrix. In *Proc. International Conference on Control, Automation, Robotics and Vision*, pp: 1242-1247.

Wermuth, N. 1976. Analogies between multiplicative models in contingency tables and covariance selection. *Biometrics*, 32, 95-108.

Williams, S. B. 2001. Efficient solutions to autonomous mapping and navigation problems. Ph.D. Thesis. Australian Centre of Field Robotics, University of Sydney.

Zhang, F. 1999. *Matrix Theory: Basic Results and Techniques*. Springer-Verlag.